BIOPUNK

CURRENT

BIOPUNK

DIY Scientists
Hack the Software of Life

Marcus Wohlsen

CURRENT

CURRENT
Published by the Penguin Group
Penguin Group (USA) Inc., 375 Hudson Street, New York, New York 10014, U.S.A.
Penguin Group (Canada), 90 Eglinton Avenue East, Suite 700, Toronto, Ontario, Canada M4P 2Y3
(a division of Pearson Penguin Canada Inc.)
Penguin Books Ltd, 80 Strand, London WC2R 0RL, England
Penguin Ireland, 25 St. Stephen's Green, Dublin 2, Ireland (a division of Penguin Books Ltd)
Penguin Books Australia Ltd, 250 Camberwell Road, Camberwell, Victoria 3124, Australia
(a division of Pearson Australia Group Pty Ltd)
Penguin Books India Pvt Ltd, 11 Community Centre, Panchsheel Park, New Delhi – 110 017, India
Penguin Group (NZ), 67 Apollo Drive, Rosedale, North Shore, 0632, New Zealand
(a division of Pearson New Zealand Ltd)
Penguin Books (South Africa) (Pty) Ltd, 24 Sturdee Avenue, Rosebank,
Johannesburg 2196, South Africa

Penguin Books Ltd, Registered Offices: 80 Strand, London WC2R 0RL, England

First published in 2011 by Current, a member of Penguin Group (USA) Inc.

10 9 8 7 6 5 4 3 2 1

LIBRARY OF CONGRESS CATALOGING IN PUBLICATION DATA
Wohlsen, Marcus.
Biopunk : DIY scientists hack the software of life / Marcus Wohlsen.
p. cm.
Includes bibliographical references and index.
ISBN 978-1-61723-002-8
1. Biotechnology—Social aspects. 2. Genetic engineering—Social aspects. 3. Amateurism. 4. Punk culture.
I. Title.
TP248.23.W65 2011
660.6—dc22 2010043689

Printed in the United States of America
Designed by Nancy Resnick
Photograph on title page and part title pages © Chad Baker/Photodisc/GettyImages

To Rosemary

Every orchid or rose . . . is the work of a dedicated and skilled breeder. . . . Now imagine what will happen when the tools of genetic engineering become accessible to these people. There will be do-it-yourself kits for gardeners who will use genetic engineering to breed new varieties of roses and orchids. Also kits for lovers of pigeons and parrots and lizards and snakes to breed new varieties of pets. Breeders of dogs and cats will have their kits too. Domesticated biotechnology, once it gets into the hands of housewives and children, will give us an explosion of diversity of new living creatures.

— Freeman Dyson, "Our Biotech Future," *The New York Review of Books,* July 19, 2007

The Commission believes that unless the world community acts decisively and with great urgency, it is more likely than not that a weapon of mass destruction will be used in a terrorist attack somewhere in the world by the end of 2013. The Commission further believes that terrorists are more likely to be able to obtain and use a biological weapon than a nuclear weapon. The Commission believes that the U.S. government needs to move more aggressively to limit the proliferation of biological weapons and reduce the prospect of a bioterror attack.

— *World at Risk: The Report of the Commission on the Prevention of WMD Proliferation and Terrorism,* December 2008

Contents

Preface

Bringing a child into the world has always taken a great measure of faith. No matter when or where, parents have always had to believe that the future offered their kids something better, or at least not something worse. Now is a weird time to have a child. I wonder if parents ever had so many reasons to be pessimistic while at the same time feeling confident that their kids have a good chance at leading remarkable lives. The promises and perils of technology weigh heavily on both sides of this emotional seesaw.

A few years ago, *The Economist* ran a cover story on some slightly obtuse advances in the scientific understanding of RNA. I knew from high school biology that RNA carried the genetic code spelled out by DNA into the part of the cell where those instructions are translated into proteins—the basic stuff of our physical selves. According to the article, these latest insights into RNA comprised "biology's Big Bang." Biology would be to this century, according to the story, what nuclear physics was to the last.

Headlines sell magazines, and the article failed to convince me that these advances in the science of RNA were biology's equivalent to splitting the atom. Yet the more I learned, the more I began to see sense in the piece's broader claim that biology—specifically, molecular biology—would be the marquee science of our time. As I paid more attention to the field, I saw the potential for transformation everywhere. Stem cells that could mimic any tissue. Scanners that could decode all three billion letters in a person's DNA alphabet in weeks. Deep insight into the genetic roots of disease. Genes hacked together to create new varieties of life not conceived by nature.

Even more than what researchers can do and do know, I marvel at how much is left to learn. Each advance in the understanding of the complexity of cells, the fundamental unit of life, brings a glimpse of an even greater complexity that scientists could not know existed until they reached the precipice of their previous knowledge. The intricate pathways from genetic code to bodily expression. The baroque folding patterns of proteins. The subtle mechanisms that tell a gene when to speak and when to remain silent. The maddening elusiveness of the cancer cell—the evolutionary equivalent of a suicide machine. Biologists have so many doors left to unlock. Perhaps the twentieth century's great push outward and upward into the sky and beyond will be matched this century by a great turn inward, into the body and its intricate secrets. We might not discover the biological equivalent of the Big Bang, but perhaps some intrepid explorer will take the biological equivalent of the first trip to the moon.

The analogy isn't perfect: The completion of the Human Genome Project in 2003 may well have been biology's one giant leap for mankind. Or perhaps that distinction could even go to the first time scientists spliced together genes from two different organisms, just four years after Neil Armstrong took his one small step. Perhaps gene splicing will be looked back on as the equivalent of the invention of the internal combustion engine and the Human Genome Project the Wright brothers' first flight. If molecular biology advances as far in the twenty-first century as twentieth-century aerospace, between Kitty Hawk and the International Space Station, life will look much different one hundred years from now.

Not long before I started this book, I took part in the riskiest genetic engineering experiment of all: I became a father. Bringing a new life into the world when life itself has become such an unstable category induces a strange kind of anxiety. Parents have always feared the unknowable future. But until the past decade or so, wondering whether your son would keep the genetic identity he was born with would have seemed a sci-fi delusion. No credible scientist would say that bioengineers are close to infusing human subjects with horse genes to run faster or fish genes for breathing under water. Yet these comic book fantasies contain a germ of credibility. If not plausible, the concept of genetically altered humans is comprehensible. The ability to tinker with our genetic essence feels not so much like a problem of basic science but one of engineering. Scientists can create a rough sketch of how to get there; they just need to discover the remaining scientific details in order to fill in the outline.

Though he won't be aware of it for a while, the epic quest now under way to trace the complete blueprint of our genetic selves will give shape to my son's life, and his entire generation, as they grow into adulthood. These children are the first to be born with the genetic map already drawn. Every advance made toward understanding what the signposts on that map mean and how they function will represent another step toward a greater understanding of a certain version of himself. Ironically, each discovery could also mean another step taken toward the means to alter that self in ways that stir both hope and horror. I want to understand what that future will look like for my son. This book is a search for that long view and the people who might point the way.

I

HACK/OPEN

The most disruptive force on the planet resides in DNA. Don't believe it? Only five days into the 2009 swine flu outbreak, the swapping of a few genes in a virus native to pigs shut down Mexico. Schools closed. Churches were shuttered. Mexico City's famed Cinco de Mayo parade was canceled.

In the United States, a few dozen cases of the flu caused by the genetically novel virus sent airline stocks tumbling. Pundits muttered about economic recovery being stopped in its tracks. Russia and China barred U.S. pork imports. Europe warned against travel to North America.

In the weeks that followed, the world discovered the new swine flu strain was not as deadly as originally feared. But its potential for sowing social and economic chaos was already clear. All because a few letters in the genetic code of the world's most primitive life forms merged, switched places, and became something new. A few bits of disrupted DNA commanded the instant attention of the world's political, economic, and social institutions with the same force as war, financial collapse, and natural disaster.

Yet plague has a way of burrowing more deeply into our nightmares than those other dark horsemen. Pathogens are invisible. They are remote from our senses, yet they are everywhere. They also have no conscience. Infection knows no morality. These pestilential microorganisms exist on an entirely inhuman scale. We feel we have no control. This scares us.

But imagine you did have control. Imagine the genetic changes that transform a harmless bit of DNA into a lethal germ were something you could see. And not only see, but understand. And not only understand, but change.

Three days after the Centers for Disease Control and Prevention first announced the detection of swine flu, Mackenzie Cowell issued this disgruntled tweet: "@CDCemergency declines to answer questions about H1N1 genome sequence identity."

Of all the things to worry about at the onset of a possible global pandemic, Cowell's slightly cryptic concern seemed low on the list. But the impish twenty-four-year-old could trust that his hundreds of Twitter followers knew what he meant and why it mattered.

Cowell is the cofounder of a group called DIYbio. Based in Cambridge, Massachusetts, Cowell and his cohort believe that biology is too important to be left in the hands of experts. By this they mean that the life sciences as practiced by academics, corporations, and the government are hamstrung by politics and bureaucracy in ways that make cumbersome the beneficial applications of the latest life-science discoveries. They also believe that computers, genetics, and engineering are fast converging toward a single point where tinkerers and hobbyists without advanced degrees will soon be able to perform sophisticated feats of genetic engineering at home.

But at the onset of the swine flu outbreak, they had a more pressing concern: A global virus-driven pandemic was breaking out, and the CDC would not release the source code.

Over the past decade, scientists have made blistering advances in decoding DNA, human and otherwise. The three billion pairs of chemicals identified by the letters A, C, T, and G that make up human DNA took the Human Genome Project $2.7 billion and thirteen years to read, an earth-shaking project finally completed in 2003. Not long after the swine flu appeared in 2009, one scientist reported that he had read all three billion letters of his DNA for $50,000. The process took a few weeks. Better computers, software, and optics are the main technological advances that have made this explosion in genetic data possible. As that avalanche of information has piled up, biologists have remarked upon the striking similarity between the code used to program computers and the genes that encode our living selves. The more geneticists learn, the more tempting it is to think of DNA as the software of life.

Members of Cowell's group and its offshoots in San Francisco, New York, and elsewhere call themselves biohackers. They cheerfully embrace the idea of human-computer commonality. If computers can be programmed, and living things are not so different from computers, they reason that life too can be programmed. Hacking in this context is not a negative concept. It does not mean breaking into systems, stealing identities, or trashing privacy. It absolutely does not mean spreading viruses. Biohacking in the form promoted by DIYbio is about engineering elegant, creative, self-reliant solutions to doing biology while relying not on institutions but wits. The solution is the hack. Hacks do not require fancy lab equipment, federal funding, or peer review. They simply need as many hands, eyes, and brains focused on a problem as possible. Somewhere in that community of creative minds, the hack is waiting. Uncovering it simply requires the access to tools, the access to knowledge, and the freedom to access both, according to the biohacker credo. When a global pandemic is looming, there can be no target riper for the hacking than the swine flu itself.

But for biohackers to take a run at hacking the virus, the code must go open source. Hence the push for the CDC to make the sequence publicly available. Let anyone who wants to take a swipe at swine flu hack away. Supporters of open-source software contend that their movement has shown the superiority of allowing creative minds to come together for a common purpose. Without the rigid restraints of ownership, they argue, ideas flourish. In the end, you get Linux, the fast, fun, free alternative to Windows. Lock up the code in the bowels of a centralized bureaucracy where ideas are compromised by the profit motive, and you get Windows Vista, Microsoft's quickly discarded upgrade to XP. Biohackers like Cowell believe that applying that same logic to pandemics could yield the quickest, smartest results, and that leaving the task to government agencies and big pharmaceutical companies means the person with the brightest idea may never get a chance.

Biohackers have some obvious backup when they scoff at the pretense of professionalism and the cult of the expert. Many of the most revolutionary computer hardware and software innovations came out of garages. Bill Hewlett and Dave Packard started their information technology behemoth in a garage. Steve Jobs and Steve Wozniak were part of the original group of hackers in the Homebrew Computer Club when they built their first Apple in the 1970s. Sergey Brin and Larry Page invented Google in a friend's garage. Mark Zuckerberg started Facebook in his dorm room. The nonprofessionals in information technology became the innovators who were then able to define professionalism on their own terms, because their ideas won out. (Suits out, black mock turtlenecks in.) In the twenty-first century, the vaccine to fight swine flu was manufactured using millions of chicken eggs, a wildly inefficient process in use for more than fifty years. Methods exist to improve upon this primitive method, but industry and government have yet to scale up the technology, even in the face of what was seen as swine flu's dire threat. Biohackers say step aside and let others try, or at least make room at the lab bench.

"It's a truism that innovation comes from putting a lot of people with different expertise in a room together and asking them to solve a problem together. The world would be a better place if we had a system or a framework that enabled passionate people who have expertise in whatever to do biology on the side," Cowell told me. "Biology's really fundamentally no different from cooking most of the time."

It's another truism that if you can figure out how to kill a pathogen, you may also be able to figure out how to make more. This is one important way in which home-brew biotech departs from basement beer making and from more traditional hacking. A cook experimenting in the kitchen can end up with a fallen soufflé. A computer builder with a soldering iron could end up with burned fingers and a useless box of metal. A biohacker who is either careless and unlucky or brilliant and evil could someday theoretically unleash a swine flu variant that resists all treatment by known antivirals and has no off switch.

The disruptive power of DNA changes the terms of the open-source argument. In computer software, knowing the source code allows a hacker to make an app that serves good or destructive ends. The terminology comes straight from biology: Malicious code is a "virus." A "contaminated" computer is "infected." Compromised computer security can create havoc. A hacked defense system in theory could send missiles skyward. But the mad scientist worst-case scenario conjured by the idea of biohacking stirs more primal fears. Tomorrow's Dr. Frankenstein would not be building a human-sized monster to stalk the villagers in plain sight. Today's most promising technologies for reading and writing DNA stir worries that he or she would be in the kitchen synthesizing a microscopic superbug no one could see even after it was too late. And maybe the blueprint for that germ would start with the genetic code for a flu virus available to anyone with an Internet connection.

Once scientists had isolated the swine flu virus shortly after the initial outbreak, it was a trivial effort to sequence the pathogen's entire genetic code. Before long, writing that code to create new viruses from scratch may become just as easy. In that light, offering that code to anyone who wants it looks as much like a path to destruction as to innovation. As the swine flu and ensuing panic demonstrated, a hacker could maximize malice by infecting the human network. Digital destruction can undermine economies and infrastructure. Viruses made from carbon rather than code can kill.

It's possible that science will never be able to create novel bacteria or viruses able to match the destructive power of pathogens evolution has perfected over billions of years. Researchers have such a hard time now defeating well-understood germs like cold and flu viruses in part because the forces of natural selection have made them strong.

New pathogens built in labs simply may not be able to compete if released into an environment already teeming with such hardy naturally occurring microorganisms.

Even so, the federal government fears terrorists will turn to biology as their weapon of choice for the next major terror attack against the United States. If that happens, the culprits probably will not even need the latest tools of genetic engineering to inflict destruction. Nature already offers up a rogue's gallery of poisons and pathogens known for their ability to maim and murder. And figuring out how to brew them up hardly takes a master's degree.

Like any powerful technology, biotech carries the potential to both help and harm. The key question in the democratization of genetic engineering is whether putting the tools and techniques of biotech into the hands of more people will tip that balance. Within the walls of major scientific institutions, biotech has yielded the most promising cancer drugs to date. Well-funded professionals are also able to create genetically augmented germs extra well suited to kill. Will more people operating with less supervision unleash biotech for better or worse? Ready or not, we may soon find out.

CHAPTER 1

Blood/Simple

"We're not good at this. I could not kill you all even if I wanted to."

For an evil genius, Kay Aull comes off as very self-effacing. In fact, the twenty-three-year-old MIT grad has no malicious intentions, unless you consider her desire to grow a tail a crime against nature. Of anyone at CodeCon, San Francisco's premier anticorporate underground hacker fest, Aull has the skills to be the most likely one who could someday slip a self-made germ into the coffee. But she says that would hardly be the most efficient way to spread a plague with any power.

Aull is tall and beanpole skinny. She carries herself with the stiff awkwardness of an adolescent boy still not used to his overlong limbs. She also speaks with the crisp confidence of someone accustomed to knowing more about her area of expertise than anyone else in the room—and she probably does. In this room, a cavernous art and performance venue called CELLspace, knowing your stuff counts for a lot. This group may share common causes—cyberprivacy, intellectual freedom, battling corporate encroachment on individual rights—but you had also better know what you're talking about. In 2009, Aull and her crew from Cambridge, Massachusetts, became the first biogeeks to take the CodeCon stage. The pressure was high: Could biotech hang with the keyboard jockeys who had already shown how a few lines of code could shake up entire economies?

Aull put herself through MIT working nights at a DNA synthesis company. Creating chunks of genes out of life's basic building blocks for profit may sound like profound, painstaking work. In practice, DNA-synthesis shops do the grunt work for research labs, which outsource the tedious mechanical work of building the same sequences over and over again to free up scientists to generate discoveries.

Synthesis companies have security in place that automatically flags orders for sequences known to be dangerous. Shops typically will not sell to customers they have not vetted. Workers have grown tired of dull journalists who think they were the first to submit an order for smallpox and can catch companies creating pathogens.

None of these protocols means that the DNA grunts could not make a contagious bug. With keys to the lab and access to the code books, Aull says she could have spent her night shifts making polio, as a few scientists did via mail order in the early 2000s, just to make a point. She did not and would not make a classic deadly germ. But she could have.

An even easier path to bioterror success would be to get a job at a federal biodefense lab, she explained. After September 11, defending the U.S. from a biological attack became a national priority. For someone with Aull's expertise, the government is always hiring. Once you got the job, you could steal a lethal microbe from the bowels of the lab and just add to water—or the U.S. mail. The FBI believes that army biodefense researcher Bruce Ivins stole from his lab to commit the anthrax attacks that terrorized a country still reeling from September 11. Ivins's alleged scheme was much easier, and much cleaner, than having complex, deadly microorganisms brimming out of beakers in a cramped apartment kitchen. Before he apparently died by his own hand in 2008 Ivins denied responsibility for the attacks.

Aull no longer works at the DNA synthesis company, and she does not work for the government. She did not lay out the route to germ-driven mass death as a how-to for the CodeCon crowd. Quite the opposite: She was pointing out that the suspicions about the work she has chosen to do are irrational. As the first MIT student to graduate from the school's new bioengineering program, Aull could have had any biotech job she wanted. Instead, she chose to set up a biology lab in her closet. She wanted to make clear to the audience that her skills notwithstanding, the road to world domination did not run through her apartment.

Still, there was something fanatical about her project. She clearly relished the challenge of doing the work at home—for cheap, relying only on her wits and creativity. She was engaging in what she felt was a more pure kind of science, a curious mind engaged passionately with nature, free of any of the most common ulterior motives: profit, career, prestige. All scientists start out as amateurs. By resurrecting her inner nine-year-old, the girl who cross-bred houseplants while her peers played Nintendo, Aull appeared to be purposely stoking her primal scientific impulse, the driving engine of discovery.

"It's not enough for me to understand how something works," Aull said. "I need to poke it with a stick."

In practice, this meant building her own gear or buying it on eBay (she bought a $10,000 cell incubator for $90, including delivery). It meant using a rice cooker and a whiskey tumbler to make distilled water ("a high-tech temperature controlled apparatus"). It meant using her cat as her chief safety officer ("If he can't play with it, I can't either"). The upside was that if she could pull her DIY wet lab off, she could do whatever project she wanted. What she wanted to do was hack some genes that could save her life.

Historians of science trace the emergence of modern biotechnology back to the first successful gene-splicing experiments in California in the early 1970s. Only two decades after the discovery of the double helix, scientists had figured out how to isolate specific sequences of DNA and insert them into the genomes of other organisms. (The manipulation of genes by humans goes back much further, to the invention of agriculture. Simply crossing wild varieties of plants and animals until they gained the docile predictability of domesticated life was the first and still the most momentous hack in biology.)

The essential feature of biotech is the deliberate mixing or reordering of genes to create something unprecedented in nature. Genetic engineering can be accomplished simply by crossing a tangerine and a grapefruit to make a tangelo. In recent decades, however, genetic engineering has come mainly to mean the process of inserting isolated segments of one organism's DNA—made up of the four chemical building blocks of all genes, represented by the letters A, C, T, and G—into longer stretches of another species' unrelated genetic material.

In pursuit of these goals, biotech has led to ever cheaper, faster, and more accurate ways to read DNA. Genomics, very roughly defined, refers to the scientific effort to decode the meaning of sequences in the genetic alphabet. Recently, advanced DNA-reading machines known as high-throughput sequencers have become a fixture in high-end biology labs. As a result, genomics has started to yield powerful medical insights.

People have always had a strong instinctive understanding that genes are the root cause of many diseases. The obsessive emphasis on ancestry and blood lineage across human cultures throughout history shows just how certain people have always been of the basic genetic truth that traits are passed on. People have not needed science to tell them the commonsense truths of inheritance: "Sickly parents have sickly children." "Weak hearts 'run in the family.'" "The men in my family tend to die young."

This last was the truth of Kay Aull's family. On her father's side, many of the Aull men never made it out of middle age. The reasons were blurry, but her dad made a point to have his health closely monitored.

When Aull's father turned sixty, his doctors noticed that his liver enzyme count was up. A high count means liver cells are being destroyed. His doctors told him to stop drinking alcohol.

Still, the numbers went higher. The doctor ordered a blood test.

As the technician drew blood, his eyes widened. Her dad's blood was as thick and viscous as pancake syrup. "This is a problem," the technician said.

The lab tech sent Aull's father back to the doctor for more tests. He told the doctor that when he moved, he felt like he had sand in his joints.

The doctor took little time connecting the thick blood to the grating joint pain. Aull's dad had an advanced case of hemochromatosis, a disease that causes the body to absorb and store too much iron. The extra iron builds up in the body's organs and causes chronic damage. The extra iron made Mr. Aull's blood thick, and iron crystals had formed in his joints. Without treatment, the disease can destroy the liver, heart, and pancreas. The skin of untreated victims turns a deep bronze. The only treatment for hemochromatosis would make a medieval physician smile. Since being diagnosed, Aull's father must give blood every month to drain the excess iron. A genetic test, available for about a decade, confirmed the disease.

Hemochromatosis is one of the most common hereditary diseases in the United States. But because its symptoms mimic so many different health problems, it is notoriously tricky to diagnose. Sufferers have died because they were being treated mistakenly for diabetes. In recent years, however, researchers have isolated the gene that regulates iron absorption in humans. Uncovering the two mutations in that gene that cause hemochromatosis was a simple next step.

When the results of the test for the mutations came back to Aull's father, her mother called her in a panic. "What does it say?" Aull asked. "I don't know," her mom said. "It's written in genetics."

Aull's mom wanted her daughter to read the results for her. "I'm going to fax you this," her mom said. "Tell me if he's going to die."

Aull's dad took a different approach. "Dad asked Google," Aull says. Aull thinks her dad responded the right way. Tools like Google put vast amounts of biological knowledge within reach, the centuries of research undertaken by the giants on whose shoulders Aull acknowledges she stands. We have more ways to know ourselves now than ever before. As biological knowledge proliferates, Aull believes, people should take responsibility for that knowing.

This article of faith compelled Aull to build her own hemochromatosis test in her apartment. The genetic test her dad took to confirm he had the disease is expensive. Those who do get tested have typically had other possibilities ruled out before they or their insurance companies will spend the money.

Aull challenged herself to build a gene test using only what was in her kitchen or what she could buy online used. She spent just over $100 on a high-voltage power supply and a table-top device the size of a bread maker called a thermal cycler that replicates chunks of DNA. An enzyme she needed for a reaction came from bacteria naturally found in hot springs. Another expense was her primer, the made-to-order DNA sequences she needed to make the test. These she obtained from the company where she used to work. This was as simple as going to the Web site, typing in her sequence's letters, and waiting for the test tube in the mail. The cost was about thirty cents per letter of DNA.

Aull's primer is designed to bind to genes containing the mutations for hemochromatosis. In her test, the primer bonding to the mutant genes would result in longer strands of DNA. If neither mutation is present, the strands are shorter. Aull tucked her lab into a closet in her small apartment in Cambridgeport, a traditionally working-class neighborhood in Cambridge downslope from MIT along the Charles River. The lab shared space with her three roommates and her cat.

Aull wanted to know if she ran the risk of contracting hemochromatosis too. She could have waited and monitored her health until she started showing signs of the disease like her father. She could have demanded and paid thousands of dollars for the standard genetic test available from sophisticated clinics. Or she could swab her cheek with a Q-tip, mix it with her primer, and stick it in the thermal cycler in the closet.

"I think the most important thing about DIYbio is it's something you can do too. It's not magic. It's chemistry," Aull said. "Doing it in the sink demystifies the process."

To separate out the different strands of DNA in her vial, Aull ran a drop of the amped-up gene samples through an electrified gel in a plastic box. The process is called gel electrophoresis, a basic tool in every college biology lab. The larger the DNA segments, the farther they will migrate toward an electrode on one side of the box. Aull ran her genes through the gel and on the CodeCon stage held up the results—the gel in a plastic Ziploc bag. Different-size DNA in gel boxes clump together in vertical bands across the gel. Large bands on the left in Aull's test meant the presence of the hemochromatosis mutation. In Aull's baggie, the long band on the left of the gel was unmistakable. The test was positive.

Aull said the mutation she carries still means that there is a less than 50 percent chance that she will contract the disease. Yet I could see her enthusiasm for the test was blunted as she talked about the results. As with genetic tests for Huntington's disease or Parkinson's, Aull could do little about what she found out other than watch and wait. She is the same twenty-three-year-old she was before the test, but now genetically haunted by a possible future over which she has little control, except for knowing to tell a doctor about her genetic predisposition if suspicious symptoms appear.

"Everyone has these deep dark genetic secrets. That's just how it is," Aull said. "Knowledge is complicated, but ignorance is not better."

Despite its ingenuity, no one would call Aull's test a biotech breakthrough, except for the drastically reduced cost. Still, the price itself reflects a deeper change in sensibility, a change spurred in part by just how cheap biotech has become. Aull's test does not represent new science but a new way of doing science. A practical piece of biotechnology based on the most sophisticated science available was built in a closet using tossed-off gear.

In a sense, technology does not transform a culture until it escapes the clutches of those who created it. That escape velocity is a function of price point. Take cell phones: In the days of battery packs carried like purses and handsets the size of large bananas, only higher-end business customers bothered to make mobile calls. In that phase of mobile technology's evolution, cell phones were luxury items, more about status than practicality. By the turn of the twenty-first century, newly cheap cell phones fueled a basic shift in expectations of everyday behavior: You ought to be able to reach anyone, anywhere, anytime. As phones became cheaper still, that expectation spread from the developed world's urban middle class to an ever-widening swath of nations, classes, and cultures.

Watching Aull describe her work, I could not help thinking back to a time my parents once told me about, when digital watches and handheld calculators cost hundreds of dollars and owning either was considered conspicuous consumption. Digital technology changed the world not only by what it could do but by how cheaply it could do it. The power of Aull's project lies not so much in what it can do but in how little it cost. Maybe her test is the biotech version of the first clumsy personal computer. Wherever her innovation sits along the PC analogy continuum, it shows the power of what can be accomplished by the biotech underground.

Self-built gene tests also offer the first hints that personalized medicine could mean much more than physicians using gene tests to make more precise diagnoses. Aull sought to learn something profound about herself. Through sheer inventiveness, she tinkered her way to that knowledge. She hacked her genes, and she gained in self-awareness. Perhaps do-it-yourself biology will someday mean a new kind of introspection: the ability to self-examine with more depth and precision than Socrates could have dreamed. (Though for now, basic biology still has a long way to go to understand other genetic variations as well as Aull's hemochromatosis mutation.)

Perhaps DIY biotech also means a new kind of freedom, where hacking your way to a greater understanding of yourself is just the first step. Synthetic biology promises the ability not just to read genes but to write them, like printing out letters on a page in a pattern that creates a picture no one has ever seen before.

In bioprophets' wildest imaginings, hacking human genes could mean making yourself into something more than human. Then again, inventing ourselves anew is the essence of individualism. Maybe giving ourselves tails or wings or chlorophyll-covered skin is just being human, fully realized, free to make ourselves into whomever or whatever we want to be. Maybe that freedom means we will not have to wait for nature anymore. This may or may not be desirable, but these are the dreams that stoke the biopunk imagination, fueled by *Blade Runner*, radical libertarianism, Newton, Darwin, and a fierce will to power and transcendence.

CHAPTER 2

Outsider Innovation

When I picture scientists at work, I still imagine white lab coats and black lab benches, even though most professional biologists I've met are more likely to wear fleece. But the second part of the stereotype is true: Labs are where science gets done. The home and the lab are separate places symbolizing separate realms of knowledge, separate practices, priorities, and precautions. This feeling holds especially true for molecular biology. Anything the size of a cell or smaller is out of scale with my domestic life, where the only cell visible to the naked eye is the chicken egg in the refrigerator.

My family does not consciously keep in our house the technology required to interact with anything the size of a cell or smaller. The only time we take time to think about life too small for us to see is when we do battle with it. Cleaners to destroy bacteria in the bathroom. Hand sanitizer to slaughter potential cold and flu. Moisturizers to shore up top layers of skin cells that betray age as they lose their elasticity. The rise of "antibacterial" as a marketing slogan on soaps exploits the widespread fear of the invisible. To deliberately bring the microscopic into our lives goes against deeply ingrained domestic instincts. And yet even as better than ever sprays, gels, and pills proliferate to defend ourselves from the unseen, some very smart people want to bring these tiny critters home.

During the first decade of the twenty-first century, a confluence of new technologies and new ways of thinking about technology virtually guaranteed that the idea of do-it-yourself biotech would take root and flower. The Human Genome Project had recently reached completion. The price and time needed to scan genes went into free fall. Meanwhile, the technology and know-how needed to combine genes into new and interesting combinations had become so easy to use that college undergraduates started building their own DNA devices.

The International Genetically Engineered Machine competition, better known as iGEM, originated out of the bioengineering program at MIT, where Professor Tom Knight developed the idea of BioBrick genetic parts. Knight started his career as a computer scientist but was inspired more than a decade ago to examine biology as if it were just another information-processing system. In 2003 he unveiled a collection of universally interlocking genetic components—DNA "parts" that each do something different but fit together the same way, like Legos.

The first iGEM competition took place at MIT in 2004 with 5 teams. The next year, 13 teams took part. In 2009, 112 teams from around the world participated. Organizers expect 180 teams in 2010. The concept of iGEM is surprisingly simple for something so technologically complex. Student teams receive a BioBrick kit—a collection of what organizers refer to as standard, interchangeable biological parts. These are strands of DNA that each perform a specific function. By combining DNA parts of the teams' choosing from the kit, students have engineered devices from blinking cells to banana-scented bacteria to arsenic biosensors. One of the competition's founding principles is a commitment to sharing and collaboration borrowed from the open-source software movement. All parts are documented in the Registry of Standard Biological Parts, an online resource that operates on the "give a penny, take a penny" principle. The registry, maintained at MIT, will send team members DNA parts that they request. Invent a new part? Add it to the registry to benefit everyone.

Mackenzie Cowell was an undergraduate at Davidson College

near Charlotte, North Carolina, when his 2005 iGEM team tried to build a genetic chemical sensor that would light up in different combinations of fluorescent colors in the presence of different substances. They didn't win, but Cowell was hooked. He moved to the Boston area after graduation and got a job in the lab that coordinated the iGEM competition. There was only one problem: Now that he was out of school, he could no longer compete. But he was not willing to let that technicality get in the way of doing biotech the way he thought it should be done.

Cowell has mutton chops and an elfin glint that have helped solidify his reputation as chief trickster among a loose cadre of do-it-yourself bioengineers in the Boston area. They call themselves DIYbio. They have T-shirts, stickers, and an e-mail list with fourteen hundred members that has become biohacking's global hub. He speaks quickly and clearly with the assurance of someone who knows what he wants to say. Yet he is not arrogant about what he believes do-it-yourself biologists can accomplish. He is probably more aware of the potential risks than anyone who might fear what he represents.

When Cowell turned twenty-five, a $20,000 trust created for him by relatives became his. Undoubtedly, they expected by that age he would have outgrown his youthful impulses and be thinking of prudent ways to invest the money for his future.

On an industrial lot in Cambridge, Massachusetts, next to a hot-tub store and an ambulance station, Cowell parked an investment in his own vision of the future and the future of biotech: a shipping container with a wet lab inside he bought at auction. He originally planned to seek donations to buy it, expecting it to cost between $6,000 and $15,000. He checked in later and was told $30,000. He scrapped his plans but went to the auction anyway. After it was over, his small trust fund was $12,500 smaller. But he now has a mobile wet lab he can take wherever someone had a big enough backyard and a good idea.

"Apparently it cost $150,000 to build, so I guess it was a pretty good deal," he said.

The Boston Open-Source Science Lab—BOSSlab—is one culmination of Cowell's DIY ambitions. Having a wet lab open to the community at large is at the core of Cowell's belief in the value of throwing open the doors of biotechnology to the most creative minds around, whatever their background or institutional affiliation. Cowell does not discount the value of all the work done by smart people handpicked by universities for their demonstrated potential and ability to contribute. But he feels certain that such a rigid filtering of minds and hands can sometimes keep tools beyond the grasp of people who should have access to them despite their lack of credentials.

Cowell's other haunt sits on a leafy street in Somerville, Massachusetts. The sprawling clapboard headquarters of Sprout & Company looks like any other house in this relaxed Boston suburb. Up the driveway, however, a garage door opens onto a workshop with some unusual extras. On one side of the room, racks of bike frames hang on the wall waiting for wheels. Machines out of high school shop class offer various ways of cutting and drilling. Upstairs houses the electronics workshop. Shelves teem with brightly colored bins holding what look like all the parts you would need to build a robot.

Across the room, inhabiting a corner next to the desktop Dell and a Peg-board draped in Christmas lights, is New England's first community-based biology workshop. The space holds everything a budding biohacker would crave. Two machines for Xeroxing DNA. A squat, steel, sterilizing autoclave, the biolab equivalent of a pressure cooker, complete with a vintage *Mad Men*–era analog pressure gauge sticking up off the lid. An electric stirring rod. Tiny centrifuges. A hulking microscope, a rack of pipettes, a deep freezer to calm squirming microbes, and a bright-orange cabinet labeled FLAMMABLE LIQUID STORAGE covered in magnetic poetry.

On a high shelf sit bottles of the wet lab's wet stuff. Distilled water. Sodium thiosulfate, a chemistry lab standby. Contact lens solution. An old vodka bottle labeled with an indecipherable chemical formula. A bottle of agarose pellets to make the nutrient-rich gelatin broth that nourishes cells and allows them to multiply in petri dishes.

The organizers of Sprout call what they do community-driven science. According to their Web site, they want to bring together people who "share their enthusiasm for investigation [to create] a kind of community college that really lives up to its name." In a region that likely has as many science labs per acre as anywhere in the world, Sprout offers an open door where everyone else demands a CV, an application fee, a grade transcript, and a card key. The deep engagement with science they promote means hands-on programs for schoolkids, thirteen-week programs for alpha geeks, and regular meetups for like-minded hackers. Philosophically, it means upholding openness as the ideal way to advance ideas.

Cowell spends a lot of time thinking about the progress of biotechnology and how DIYbio fits in. He sees several trends feeding into one another to bring molecular biology to the masses.

The first is the rise of the wet lab as a service industry. Right now, hobbyists willing to spend a few hundred dollars can mail off a swab from a kitchen sponge or a vial of spit to a sequencing company and receive back by e-mail a complete scan of the DNA found in the sample. The Web sites of DNA synthesis companies let online visitors type their combinations of A, T, C, and G right on the home page, give their credit card numbers, and get sequences made from scratch delivered by FedEx. As once complex, experimental lab tasks become mundane, more companies may emerge to take on the busywork burden. As competition drives down prices, Cowell believes researchers will no longer be limited by their time or gear. Instead, he hopes ideas and inspiration will become the only thresholds to scientific accomplishment. Biohackers will not only be freed from the need for ties to well-heeled institutional labs but from any wet lab work at all, if they wish.

"They can sit at home without any infrastructure anywhere in the world and cause experiments to be done," Cowell told me.

For biohackers who like their labs wet, Cowell believes the DIYbio approach will put the tools they need within reach. "Low-cost, simple tools and techniques built by the amateur community can lower the infrastructure costs by a factor of ten or a hundred."

Another advance that excites Cowell and mainstream scientists alike is a technology called microfluidics. Microfluidic devices shunt liquids through microscopic channels much like electricity through a transistor. The best-known use of microfluidics is the inkjet printer. At Harvard, acclaimed chemist George Whitesides has created biological microchips, though instead of silicon, these chips are made out of paper. Whitesides and other researchers are using microfluidics to build so-called labs on a chip. But Whitesides is after something more profound.

On microfluidic chips, liquids encounter other substances that are lodged in the chips' microscopic channels. The resulting reactions can tell researchers something about the nature of the substance. Blood, for instance, could pass by human antibodies that fight off a specific disease. If those antibodies react, that means the blood has tested positive for the disease. Other devices being developed can track white blood cells from the immune system in a drop of blood. In the presence of biological weapons, the number of white blood cells would jump—an early-warning biosensor.

In the early 1990s, the first microfluidic devices were etched on silicon and glass. They used electricity and pressure to move liquids through their channels. On Whitesides's paper chips, fluids move through channels like a paper towel sopping up spilled coffee. Whitesides improves on the paper towel and silicon both by imprinting his chips with a special water-repellent chemical that drives anything water-based through particular channels and toward target chemicals. When the liquid reaches its destination, the paper changes color, like a home pregnancy test. Whitesides believes the technology will bring cheap diagnostic tests to the developing world and has started a nonprofit to make that happen. Mackenzie Cowell believes that Whitesides's device could make starting a biology lab as easy as buying a $10 chip and a few bottles of chemicals.

The DIYbio crew seeks to make science more playful, and bringing down the price is part of their effort to make that happen. Another part is place. They believe science can happen in a garage. Or in a bar. Important work gets done in buttoned-up fashion in buttoned-up labs. But why not imagine science another way, they say, an approach that retains the personal spirit of the scientists involved? Why not see what happens when science is done in style? This approach is not merely an attempt to project a little hacker chic, although fashion does play a part in the appeal. Playfulness, fashion, and a direct appeal to fun all play a part in an attempt to use style to draw people toward the substance of science.

This unorthodox approach to something as traditionally buttoned down as biology catches many people off-guard. Charlie Schick, a Finnish biohacker based in Boston, described in a blog post the problem of explaining to his father what exactly he does.

"Being a biz guy, he kept asking me what was the 'end goal' to help him wrap his head around what would motivate folks to tinker with biology," Schick wrote. "He wanted to know if there was a scientific goal or if there were products folks wanted to build."

Schick, who goes by the Twitter handle @molecularist, did not have a clear answer. The talk went round and round, he wrote. Finally, Schick's father hit upon a phrase that was as modest in describing the movement's short-term goals as it was bold in predicting its long-term implications. He told Schick that he and his friends were trying to "increase the tinkerability" of biology.

Noninstitutional biologists are not on the brink of major scientific breakthroughs as conventionally measured. They are not about to cure cancer when an eleven-thousand-employee, $80 billion company like Genentech has so far failed. They are not going to unleash the world's first artificial amoeba tomorrow or graft wings onto house cats.

What they are doing is something more subtle, Schick implies, something that could make all those scenarios more plausible sooner than if biotech were left solely to the more rigid institutional patterns of innovation. Biohacking is about what Schick calls "simplifying and domesticating" biology.

And that loosening of codes would not in the biohacker vision serve only those already in the game. Making biotech more user-friendly—whether technically, financially, or aesthetically—will ideally make it more engaging for people far outside science's traditional institutional boundaries, Cowell says. Outsiders should be shown that they can have fun with science, can play with these tools and ideas that are treated with such gravitas, and that in doing so, they may begin to feel closer to themselves as they unlock the mysteries of their own organic being.

"Those people on the fringe without an agenda doing it for themselves . . . I expect every now and then, they [will] have a great idea that really works," Cowell said. In science as in most things, he says, he subscribes to the ideal of "let a thousand flowers bloom."

"The more actors you have, the more innovation can occur."

Bryan Bishop was sitting in Sprout with a mostly eaten bag of tortilla chips, salsa, and empty bottles of off-brand soda. At twenty years old, Bishop was technically too young to drink at the beer party he was hosting. But unlike most people his age, drinking seemed low on his list of interests anyway. On this muggy June night he had other things on his mind, like what to do with the industrial-strength robotic arm he had just bought off Craigslist.

"Check it out," he said, showing off a smartphone photo of the arm in the bed of a pickup truck like a proud dad showing off pictures of his new baby. "We drove around with it in the back of the truck for a while just to mess with people. Do you know what kind of looks you get when you're hauling around a robotic arm?"

The arm was the latest prize in Bishop's quest to build a first-rate hacker space in his hometown of Austin. He had a four-thousand-square-foot space, he said, about a mile from home, close enough for him to ride his bike. From the sound of it, however, Bishop did not have much interest in spending time at home. Not when he could be playing with the factory-grade laser cutter he had also found on Craigslist to trick out his new space.

On the DIYbio list, Bishop is a prolific source of insight and opinion, links and transcripts. He is also the youngest of the core group of biohackers who have ventured beyond the idea stage and are trying to back their enthusiasm with lab work. (In one of his best known hacks, Bishop used little more than a piece of paper, two sheets of glass, and a Sharpie marker to create a primitive microfluidics device.) His self-assurance, his easy command of scientific jargon, and his ability to articulate his interests and ideas command his peers' respect. While many do-it-yourself biologists have Ivy League degrees, Bishop dropped out of the University of Texas as a sophomore. He needed time to pursue "other opportunities," he said.

Bishop is a little cagey about what all those opportunities might be. He would not say, for instance, where the money came from to buy the robotic arm, which he says retails for $2 million. His funder wanted to keep a low profile. Because he bought it used on Craigslist, he paid considerably less. But he said he couldn't say how much.

Yet he has no problem talking about his goal, which is why he's hosting this party. Bishop had come to town for an event called the Humanity-Plus Summit, H+ for short, the latest incarnation of a movement known since the 1980s as transhumanism. As Bishop himself described it, transhumanism broadly defined is focused on "human advancement." In practice, this means a near obsession with the promise of technology not only to solve the most basic human challenges but also to strip away human limitations. According to transhumanist orthodoxy, aging is an engineering problem to be solved. The brain is merely the world's most advanced computer. Fusing organic gray matter with digital processing power should require just a few more decades of research at most, according to the transhumanist faithful. When that happens, expanding your brain power would take no more effort than upgrading the RAM on your laptop. Similarly, consciousness amounts to a collection of algorithms. With the right interface between flesh and silicon, the right brain-imaging equipment, and a clear map of the brain's billions of synaptic connections, downloading your mind—which transhumanists equate with yourself—to a hard drive or uploading it to the Internet becomes a mere matter of having enough digital storage space and bandwidth. Bishop described his goal as creating a hacker space in the form of a "transhumanist tech co-op." He wants to be the guy who makes transhumanist dreams real.

Right now he still has far to go. But he has no doubt that the one sure way not to get there is to waste time asking other people for permission to try. A strong libertarian streak runs through the transhumanism movement, earning it both liberal and conservative critics who chastise its followers for ignoring or failing to see the moral and political implications of their techno-utopian ambitions. Certainly the speakers at the summit were more interested in promoting their idealized vision of the future than mulling how their goals fit into a broader historical or cultural context. Their bias was toward action.

"You should go out and build the transhumanist dream," he urged the audience during his summit talk. "Join the narrative. Make stuff."

Biohackers want to make the rest of us okay with the counterintuitive coming together of biotech and basements, of DNA and dinner tables. They not only believe the mass migration of biotech out of the lab and into the home should happen. They believe it will happen soon and that we all should pretty much not worry. But the sell may not be easy.

"When we happen upon a technology such as stem cell regenerative therapy, we experience hope," economist W. Brian Arthur wrote in his 2009 book, *The Nature of Technology*. "But we also immediately ask how natural this technology is. And so we are caught between two huge and unconscious forces: Our deepest hope as humans lies in technology; but our deepest trust lies in nature. These forces are like tectonic plates grinding inexorably into each other in one long, slow collision."

CHAPTER 3

Amateurish

Advances in the understanding of human biology came slowly at first. Physicians of ancient Egypt, Greece, and Rome each contributed to a growing body of knowledge, as did the medical doctors of medieval Arab and European societies. Yet even the father of modern medicine, Hippocrates, subscribed to entirely fanciful ideas about the nature of physiology and disease. His notion that all sickness was caused by the "vapors" and imbalances among blood, black bile, yellow bile, and phlegm, aka the four "humors," persisted for more than two thousand years, leading to sadistic, useless (except in the case of hemochromatosis) practices, such as bleeding.

Perhaps no invention ultimately did more to bury humorism in the graveyard of bad ideas than microscopy. The invention of modern optics in the fifteenth and sixteenth centuries did for Western medicine what the journeys of the European explorers to the Americas did for Western culture. The microscope opened the scientific imagination to the radical possibility of an utterly new world existing just out of sight. The history of medicine is threaded with ideas about invisible forces at work on the body. After the microscope, scientists could begin to confirm that such forces were not supernatural but just impossible to see without better lenses.

In 1665, the thirty-year-old English scientist Robert Hooke published *Micrographia: Or Some Physiological Descriptions of Minute Bodies Made by Magnifying Glasses with Observations and Inquiries Thereupon*. The gorgeous volume includes intricate illustrations of everyday organisms seen as never before: a fly's eyes and wings, a bee's stinger, feathers, frost, seaweed, poppy seeds, and, most famously, a flea. (Scientists would not discover until more than two centuries later that fleas were responsible for transmitting bubonic plague, the great plague ravaging London the same year the *Micrographia* was published.) Hooke writes of the flea with a palpable sense of wonder: "But, as for the beauty of it, the Microscope manifests it to be all over adorn'd with a curiously polish'd suit of sable Armour, neatly jointed, and beset with multitudes of sharp pinns, shap'd almost like Porcupine's Quills."

Hooke's most remarkable observation lacks the baroque beauty of his insect drawings. "I took a good clear piece of Cork, and with a Pen-knife sharpen'd as keen as a Razor, I cut a piece of it off, and thereby left the surface of it exceeding smooth," he writes. Under the microscope, Hooke observed that the cork appeared to consist of "pores, or cells" made up of very thin walls surrounding otherwise empty chambers. The cells were empty because cork is a nonliving tissue. Still, Hooke had seen for the first time the basic unit that makes up all life. He did not know what he was seeing, but he knew he was the first: "indeed the first microscopical pores I ever saw, and perhaps, that were ever seen, for I had not met with any Writer or Person, that had made any mention of them before this."

Prior to Hooke's usage, the word "cell" referred to the small, bare rooms inhabited by monks. In the two hundred years following Hooke's drawings of cork, cell theory would emerge as a basic principle of modern biology. Yet the connection between cells and genes remained hidden until one monk stepped out of his own cell and into the garden.

Every high school biology student learns the story of Gregor Mendel, the Austrian monk whose experiments with peas led to the first clear understanding of how offspring inherit traits from their parents.

Mendel's status as scientific folk hero stems in part from just how far he stood outside the scientific establishment of his day, much like Einstein at the patent office. Yet the sentimental version of Mendel undermines an appreciation of his achievement's magnificent unlikeliness.

Mendel was born in 1822, the son of a farmer near a small Prussian village in what is now Poland. He showed academic promise early on and dreamed of great things for himself. But he turned gloomy and anxious as a young man as he struggled to put himself through university. Even with a gift of unexpected financial support from his sister, he could not manage the cost of a true higher education. "That is when he chose the only path available in nineteenth-century Mitteleuropa for a penniless young man in search of an education," wrote Robin Marantz Henig in her biography *The Monk in the Garden*. "At the urging of his physics professor . . . who was also a priest, Mendel signed on with the monks."

Fortunately for Mendel, the Augustinian monastery in Brünn (now the Czech city of Brno) did not lack for scholarly opportunity, though neither was it a hub of nineteenth-century European intellectual life, at least not compared to Vienna, a full morning's journey away by train. The abbot encouraged scientific curiosity among his monks and was supportive of Mendel's apparent gifts as an amateur naturalist. Mendel also turned out to be a terrible priest. The abbot sent him to teach science at a secondary school instead.

Though admired by his fellow teachers, biographers believe Mendel suffered from severe test-taking anxiety. He twice failed the exam to become certified as a high school science teacher, though after the first failure at least one examiner believed Mendel's main problem was that he lacked a real university education. He managed to hang on for two years at the University of Vienna, where he finally received training in the rigors of experimental science and data analysis.

Back at the monastery, Mendel devoted himself to the quiet study of his ever-expanding garden. Scholars debate whether Mendel fully grasped the significance of what he was learning through his investigations. But he intuited like none had before him that math would lead toward the solution to the puzzle of heredity in a way that microscopes at the time could not.

Through simple, careful, patient observation and documentation, Mendel over seven years would find that his fabled peas would not inherit blended traits from their parents. The offspring of a coupling between a yellow pea plant and a green pea plant would never yield yellow-green peas, but only yellow *or* green. And he observed that in large enough populations those traits would appear in easy to predict 3-to-1 ratios. Mendel did not know why this was so, but researchers over the next one hundred years would confirm the biological basis of the dominant-recessive theory of inheritance that defines modern genetics. Mendel demonstrated conclusively that parental traits do not blend in offspring. Instead, if either parent passes down a dominant gene, the child will express that trait (brown eyes, for example). If neither parent passes down the dominant gene, the child will express the recessive trait (such as blue eyes). Mendel showed that over time, the ratio of dominant to recessive will always work out to about 3 to 1.

Mendel read his paper detailing his results before the Natural History Society of Brünn in 1865 and in 1866 published his results in the *Proceedings of the Natural History Society of Brünn*, a scientific journal remembered to history perhaps solely for Mendel's paper. In his lifetime he remained nearly as obscure as his findings were revolutionary. His paper did not draw widespread interest until three scientists reproduced his results independently of one another at the turn of the twentieth century, years after Mendel's death. Only then did biologists recognize that an amateur horticulturalist had beaten his contemporaries to the truth before scientists knew the race had even started.

Mendel never rose as a professional scientist above the rank of "uncertified substitute teacher." Yet as his story shows, the hack's the thing. His hunch that statistics might yield an answer to the question of inheritance led an Austrian monk with no official credentials to an insight that eluded as towering a figure as Charles Darwin, who had published his theory of natural selection in *On the Origin of Species* in 1859 and was so lauded and influential in his own lifetime that he was one of only five nonroyal British subjects during the nineteenth century to receive a state funeral. Meanwhile, Mendel's monastery burned most of his papers after his death because of a dispute over taxes in which he had become embroiled while serving as abbot.

Yet his legacy persists, not on the merits of his professional stature but on the power of his idea. Mendel did not need a PhD to succeed. It was enough that he was a geek.

By the time of Mendel's posthumous acceptance into the scientific establishment, scientists had isolated DNA and observed chromosomes in cells, though they had little understanding of what either signified. As the twentieth century progressed, pioneering genetics research no longer took place in the monastery gardens of the Habsburg empire but in labs at British and American universities. Thomas Hunt Morgan in 1910 began work at Columbia University, which he continued later at Cal Tech, that pioneered the now ubiquitous use of fruit flies in genetics research to uncover the arrangements of genes on chromosomes and their role in heredity. Others refined this understanding, including with the discovery that DNA and not proteins contained the genetic material passed down from parents to offspring.

Despite their academic affilation, UCLA anthropologist and science historian Chris Kelty says Morgan and his "fly-boys" were hackers. Kelty writes:

> They built an extensive global network of fly geneticists who shared the information and the mutants via a kind of proto-internet newsletter. . . . They stitched together a research agenda out of an easily found species, and a

> bunch of ad hoc, cobbled-together tools. Their newsletter is full of clever suggestions for hacking flies: fly-food recipes, techniques for counting flies, tools to keep them warm, or cold, suggestions about how to pursue one line of research or another. They were artisans and craftspeople of science, but ones willing to standardize and extend their practices, to incorporate other fly geneticists and eventually to dominate the field of genetics.

Kelty argues that Morgan and his cohort followed a typical American trajectory: the elite outsider who becomes the insider. Nowhere does that narrative play out more famously than in the tale of the two geeks who goofed their way to the discovery of the true structure of DNA.

When I picture James Watson, he always looks like Jeff Goldblum. As a junior in high school, I spent much of my summer at Carnegie-Mellon University in Pittsburgh, Pennsylvania, at a state-run science program for the flagrantly nerdy. The most vivid memory I have of the biology course aside from bending over a microscope in mild incomprehension was the showing of the 1987 made-for-TV movie *Race for the Double Helix* starring Goldblum as Watson, one half of the duo responsible for the twentieth century's most brilliant act of scientific savoir faire. Whoever cast him in the role must also have seen that the Goldblumian glint epitomizes the special slickness with which Watson and his coconspirator, Francis Crick, pulled off the discovery of the double-helix structure of DNA. Though the Nobel committee did not say so in its 1962 citation to the pair, Watson and Crick pulled off one fantastic hack.

The two were hardly amateurs when they met at Cambridge University in 1951. Though still in his early twenties, Watson was a scientific prodigy who had earned his PhD in zoology by age twenty-two. Crick, a physicist, was more than a decade older and a slacker by comparison; he had not finished his thesis by his thirties, though in fairness he was likely sidetracked by his stint designing magnetic mines for the British admiralty during World War II. Still, he had spent most of his adult life in the halls of high-end academia working as a professional scientist. The pair were scientific insiders with access to all the resources of one of the world's leading universities.

Yet the story of their discovery reads like the opposite of the romantic stereotype of the genius scientist working feverishly, alone in the lab, hunched over bubbling beakers in the light of a full moon. In fact, neither Watson nor Crick did the actual lab work that led to their breakthrough.

The state-of-the-art research being done at the time involved a technique known as X-ray crystallography. Scientists at top labs were processing organic molecules, mostly proteins, to form crystals, which they then bombarded with X-rays. Photosensitive plates recorded the distinct diffraction patterns cast by the radiation after passing through the different kinds of molecules. Researchers believed they could use these patterns to gain insight into the basic structures of biological substances and so better understand how they worked.

Watson and Crick believed that these patterns could reveal the structure of DNA, but they did not have the skill to create the plates themselves. Instead, in one of the most famous double-dealings in the history of science, a colleague and rival of Rosalind Franklin, passed the young, brilliant scientist's unpublished images of DNA's crystalline structure—the clearest ever obtained—to Watson and Crick—without her knowledge. The images confirmed their hunch that DNA was a double helix. They relied on the biochemical experiments of others to piece together the various components of the molecule, from the sugar-phosphate backbone to the base pairs they realized encoded the genetic blueprints for all living things. In short, their "discovery" turned out to be a work of synthesis: taking the pieces of knowledge already available and rearranging them until the answer made unimpeachable sense. Their single-page paper, published in *Nature* in 1953, testified to the certainty of their solution. The double helix required no elaborate explanations. Its elegant simplicity made the design's accuracy self-evident.

In *A History of Molecular Biology,* Michel Morange disagreed with the notion that "Watson and Crick inaugurated a new style of research in which discussions and theories became more important than experiments and observations." Someone had to do the painstaking work in crystallography and biochemistry without which the pair could not have built their conclusions. Still, Morange conceded that they discovered the double helix without doing a single experiment on DNA themselves: "A close study of Watson and Crick's approach . . . reveals that their research . . . was more akin to tinkering than the work of an engineer or an architect."

Yet their tinkering was far from innocent playing around. Their use of Franklin's data and the way they obtained it left a bad taste in some mouths, compounded by the scant credit she received, even though her paper detailing her observations was published alongside Watson and Crick's famous account. In the years following Franklin's death from ovarian cancer at age thirty-seven, she became an icon for those campaigning against sexism in the sciences. In *Race for the Double Helix,* she was portrayed as the third hero, on the verge of unlocking DNA's secret herself but stymied by the unfair obstacles that kept brilliant women from achieving as much as their male colleagues. The atmosphere of suspicion and infighting that had both kept her data behind closed doors and led to its unauthorized release typified an institutional culture that biopunks and many professional biologists agree still afflicts the practice of biology today. Had a culture of openness and data sharing existed at the time, perhaps Franklin's images would have been available already. No one would have needed or could have had a way to betray her by passing along her data without her knowledge, nor could her contributions to the discovery have been downplayed, since as part of the intellectual commons they already would have been widely known.

Watson and Crick's hack was perhaps as much about working the system as the science. They cleverly navigated scientific institutions to get the knowledge they needed. Biopunks might argue that freeing scientific knowledge from such institutions would let anyone with the intellect and insight become an armchair Watson or Crick. And the shedding of proprietary ownership of knowledge would also eliminate secrets. Greater transparency would mean credit would go where credit was due. Future Watsons and Cricks could tinker without venturing into an ethical fog of credit and recrimination.

CHAPTER 4

Make/Do

In *Blade Runner*, synthetic humans shop seedy storefronts for new body parts. In an apartment in San Francisco just before Christmas, self-taught bioengineer Meredith Patterson spliced genes at her dining room table. Patterson was not making a "replicant," at least not yet. But her home lab felt like the place where, centuries from now, a genetically engineered android might look back and say, "That's where it all began."

In Patterson's apartment, stacks of dirty dishes teetered in the sink and on the counters. Cinder-block shelves sagged with science fiction, computer programming tomes, and books about cryptography. Cats sauntered among boxes spilling over with the guts of half-dismantled electronics. Above the stove, spices of every color and kind were packed into identical, clear canisters evenly spaced on a small shelf in a perfect zigzag, a jarring display of order amid the cheerful disarray.

Genetic engineering is delicate work. The tasks are repetitive and time consuming. They can strain the patience of anyone who is less than passionate about decoding what Francis Collins, the head of the Human Genome Project, called "the language of God." Much of your time is spent making sure you do not kill your "bugs," the microbes containing the DNA you are trying to tweak. Occasionally you work with bugs that you need to make sure do not kill you.

At her table, Patterson plunged a large, narrow eyedropper into a large glass beaker of clear liquid. The tattooed thirty-one-year-old sucked a few drops into the tube, squeezed the liquid into a small plastic vial, closed the lid, and jammed the two-inch container into a small piece of Styrofoam alongside several dozen identical vials. She repeated the process over and over again.

Patterson is a self-taught computer programmer, a linguist, a sci-fi author, blogger, knitter, and handgun aficionado. She carries a stack of ideas teetering in her head at any one time. Her brain cannot stop collecting, consuming, taking things apart, and reassembling. But at the center of the swirl an intense ability, even a need, for analysis and organization takes hold. When Patterson encounters a technology for the first time, she does not just absorb the general shape. She goes straight for the details. She feeds on the logic of the technical. When she does, she can speak and write with great precision about what she's learned. And then she gets to work.

A self-described "gun-toting liberal" humanist with a libertarian's passion for self-reliance, Patterson told me she would like to live in a "do-ocracy": a society where people have the knowledge and the means to build anything they need for themselves. Her kitchen-table biotech project was an attempt to spread those means to a huge swath of people recently victimized because they had no good way to protect themselves. With her beakers and vials, Patterson hoped to create a cheap, decentralized way to secure the milk supply of the developing world against unscrupulous dairies that were using poison to bulk up profits.

As Patterson hunched over her setup, China was embroiled in scandal. Tainted infant formula had killed four infants and sickened hundreds of thousands. Because of the illnesses, the world learned for the first time that Chinese dairy producers had been cutting their milk for years with a chemical called melamine. Properly used for making plastics and cement, melamine can also fool simple tests to make dairy products appear to contain more protein than they actually do. A Chinese court ultimately sentenced a cattle farmer and a milk trader to death for this practice.

Yet even as the sources of the contamination were traced, the world was in a panic. The tainted formula had traveled far and wide in China and, many feared, beyond its borders. Compounding these fears was the revelation that this practice was not new to Chinese milk producers. A tainted international baby food supply seemed not just plausible but likely. Still, no one knew for sure.

Patterson does not necessarily distrust Food and Drug Administration claims that the U.S. food supply is safe. But she does not see why she, or anyone else, should have to wait around to be told. (The FDA's own testing would reveal in November 2008 that melamine was in fact present in low levels in U.S. infant formula.) She wants to be able to find out for herself and wants anyone else to be able to do the same.

The standard tests for melamine in food require lab equipment that costs thousands of dollars. Assuming you had the money, the inclination, and the room in your home for the sprawling machinery, parents who wanted to protect their kids would still need advanced degrees to make it work. This scenario violates every techno-aesthetic standard a hacker holds sacred. It's cumbersome, esoteric, and top-down. This was why Patterson was in her dining room working on a test that she hoped would cost families no more than one dollar and be as easy to use as putting a few drops of milk on your wrist. Her not-so-simple plan: splice a glow-in-the-dark jellyfish gene into the bacteria that turns milk into yogurt and add a biochemical sensor that detects melamine. Yet the complicated device, if it worked, would be simple to use. To stay safe, a rural family in China or Cameroon or Kansas would just have to mix their milk with Patterson's custom genes and make sure the combination did not turn green.

"The really cool ideas that end up making the world a better place, that end up curing diseases and making the world healthier, often don't come from the corporate establishment. If we have to wait for university labs to do this, we're going to be waiting forever," Patterson told me during a midexperiment smoke break.

"As I think the open-source software world has shown us, innovation comes from people seeing there's a problem and deciding they're going to figure out how to solve it. If the kind of innovative person who gets into open-source software sees some kind of biological problem and says, Hey, I can figure out how to solve that, we can tap into the same kind of innovation that brought us the Internet, that brought us Web 2.0."

If one must earn the title of hacker rather than simply claim it, Patterson's primal urge to tinker fulfills one of the prerequisites. In the hands of its most gifted practitioners, tinkering is an essential form of creativity. But it is a different brand of creativity, practiced in a different spirit, than the kind suggested by the romantic image of the lone artist or genius inventor trying to wrestle inspiration out of nothing.

The idea of technological innovation as a function of inspiration has loomed over Western culture since Prometheus stole fire from the gods. In the ancient Greek myth, humanity's champion among the immortals rebels against the cosmic order and in one bold stroke delivers us our most essential technology. According to the Promethean model, innovation is achieved by brave, isolated individuals storming Mount Olympus, where knowledge is shrouded in great clouds of mystery and peril. Those who make it back are hailed as heroes and often make a great sacrifice for their efforts. (Zeus chained Prometheus to a rock to have his liver gnawed out daily by an eagle; Marie Curie pioneered the study of radioactivity and died of radiation poisoning.) The history of science and technology has traditionally been presented as the heroic march of great (mostly) men and their works of genius. In the twenty-first century the theater of the innovator as Promethean hero plays out every six months or so, as Steve Jobs takes the stage in San Francisco to bring us the latest spark snatched from the bowels of Mount Cupertino.

Hackers do not much go in for hero worship. This is partly driven by the anonymity that is often an occupational necessity when skirting the margins of legality. The hack is the thing. And if you gain a little name recognition, the next question is, What have you hacked for me lately? Cleverness and quickness are valued over any suspect notions of genius. In this view, Steve Jobs is more P. T. Barnum than Albert Einstein. As such, hackers place little stock in the Promethean vision of creativity. Their innovators are not heroes. They are mischief makers. And these tricksters do not sweat their way to discovery. They tinker.

The figure of the tinker has a rich, complicated history in Irish and Scottish culture. Today the term is considered a slur against indigenous minorities known properly as Travellers in both countries. As the name suggests, these groups practice a nomadic lifestyle similar to the Roma people, or Gypsies, of continental Europe, though they are not ethnically related. Historically, the term "tinker" as a noun means tinsmith and refers to Travellers who made a living going from town to town to mend residents' pots and kettles. For centuries the word was also used more broadly to describe anyone at the bottom of the social hierarchy.

As such, to tinker was to be a vagrant. Tinkering was not the work of an active, contributing member of society. The tinkerer was an outsider, not to be trusted. Over time the tinkerer came to be portrayed, in Irish literature especially, as a trickster, a rogue, the canny clown who played the fool but was able to get his way with the clueless sedentary people who saw themselves as socially superior. That image persists as a crude stereotype of shifty grifters inflicted on today's Travellers (think Brad Pitt in *Snatch*).

Tinkering has long since taken on a more generic sense of fiddling or tweaking, of spending Saturday afternoon in the garage trying to squeeze a few more horsepower out of the bitchin' Camaro. But it still retains the idea of work that is not really work. Jacking up your shocks and putting balloon tires on your F150 pickup truck is not something you do because you have to. Tinkering is work you do for fun.

Hackers like Patterson have embraced the playfulness of tinkering, but here's the mischief in their creed: Just because the work is fun does not mean it is unimportant. "Playing," in the hacker sense of the word, is not just a way to stay entertained. It is an attitude toward innovation that champions gamesmanship, that prizes intellect applied with competitive vigor and flair. In chess, the grandmaster and the goat each play with the same sixteen pieces. But in the hands of the former, the game becomes an object of beauty and raw intellectual force. In the same way, the gifted tinkerer can rearrange the already existing engine parts or snippets of computer code in a way that creates something utterly new and potentially transformative. And despite everything you have ever learned about the Protestant work ethic, the process does not have to be painful. No one had his liver gnawed out. Maybe it was even fun.

The connection between biology and tinkering is also not an entirely new notion. The Nobel Prize–winning French geneticist François Jacob in the 1970s famously personified evolution itself as a tinkerer. Israeli biologist Uri Alon elaborated on Jacob's ideas in a 2003 paper published in *Science* that describes how evolution pushes organisms to function more like engineered systems. The key to evolutionary success is adaptation. Organisms adapt more easily when they can simply reconfigure the pieces they already have in place, Alon says. A bird or a beast that had to evolve from scratch could never keep up with the creatures already thriving that would only need a few pieces rearranged to remain the fittest. "Rather than planning structures in advance and drawing up blueprints (as an engineer would), evolution as a tinkerer works with odds and ends, assembling interactions until they are good enough to work," Alon writes. "It is therefore wondrous that the solutions found by evolution have much in common with good engineering design."

Before Prometheus was ever a twinkle in some fireside storyteller's eye, our barely bipedal ancestors on the African savannah did not spend much time considering distant gods on the mountain summit or in the sky. For them, the supernatural and the natural were identical. Every rock, bone, and stick had magic in it, so much so that the magical would have been indistinguishable from the mundane. One day, as *Homo habilis* sat bored and hungry beside a creek, he or she was throwing rocks and watching the splashes they made in the water. After a while, a prehistoric pig or the original chicken came by to get a drink. That ancestor of us all, with a slight rumble in the belly but really just for fun, took one of these rocks and winged it at the beast, knocking it square between the eyes. *Homo habilis* now had dinner. More important, *Homo habilis* now had a tool—and all because of a goof. If you believe that using tools is the essential feature that separates us from the other animals, then tinkering may be the most human urge of all—a truly primal instinct. If so, the impulse to biohack starts to seem self-explanatory. Since the emergence of biotechnology, critics have often asked, Why would someone want to tinker with biology? But in light of the human need to fiddle, perhaps the more urgent question is, Why would someone not want to?

In January 2010, Patterson flew into Los Angeles from her new home in Belgium, where her husband was earning his PhD in cryptography, to call biohackers to arms. The occasion was the first academic conference devoted to what organizer Chris Kelty had decided to call Outlaw Biology? (the question mark at the end is deliberate). It was a strange coming together of anthropologists, sociologists, and the biopunks they were proposing to study. Meanwhile, the name itself had riled some do-it-yourselfers, who felt that advertising themselves as outlaws was an unwise invitation to a backlash. Patterson was not among the handwringers. In an October 27, 2009, post to the DIYbio.org mailing list prior to the conference, Patterson wrote: "I have every intention of sticking it to the man before the man sticks it to me."

Dressed in her customary black leather trench coat and biker boots, Patterson stood before the crowd and laid out her terms.

"I suppose you could call me a hacker, though that title is really more earned rather than taken," she said. She described her background—linguistics, data mining, computer security—and said that all three had something in common with biology: looking for patterns. She put up a chart of the biochemical pathways of *Homo sapiens*—all the ways the chemicals in our bodies react with one another to accomplish the baroque physical processes most of us lump under "living." Patterson had another name for it: spaghetti code, an insult programmers use to describe sloppy work. Patterson said she could not help but dream of how to tweak the tangle: "When I look at that I think, Wow, this is cool; what can I do with that?"

On that note, Patterson debuted "A Biopunk Manifesto," a call for the right to research and a vow to put the tools to do science into the hands of everyone.

"Scientific literacy is necessary for a functioning society in the modern age," Patterson said. "Scientific literacy is not science education. A person educated in science can understand science; a scientifically literate person can *do* science."

The ability to do science has a liberating effect, she said. Unlike many scientists, she does not shy away from the political implications of her work.

"Scientific literacy empowers everyone who possesses it to be active contributors to their own health care, the quality of their food, water, and air, their very interactions with their own bodies and the complex world around them."

Yet as much as Patterson was calling biohackers to the barricades, another impulse also animates her ambitions for biology. In conversation, Patterson is always serious, and she is defiantly articulate. She dares you to question whether she knows what she's talking about. At the same time, she cannot suppress her geek's glint—the sparkle exuded by the avowedly nerdy when discussing what they truly love.

Patterson first learned to play around with biology at a DNA synthesis start-up, where she went to work not as a trained biologist but as a database engineer, a field she came to in her typically eclectic fashion, via linguistics. Patterson grew up in the Houston area. Her father was a chemical engineer. He brought home an IBM PC Jr. when she was eight years old and helped her learn to program. Doing things yourself was big in her house growing up. "He was always there to answer questions when I had them," she says of her dad. She got into

Usenet during the early Internet days, a mark of authentic geekhood for nerds of a certain age. She took every science class she could but also wrote a lot. In college, she studied creative writing and started taking upper-level linguistics classes. The writing did not stick but the linguistics did: computational linguistics led to databases led to bioinformatics—the science of processing and interpreting the deluge of data generated by biological research.

She started a PhD in computer science and began writing software for Integrated DNA Technologies, a cutting-edge gene maker improbably located in Iowa, where Patterson was going to school. The company itself had a solid DIY ethic, she says: When they didn't like a piece of equipment, they tried to make a better, cheaper version themselves (a salad spinner centrifuge, a gene-copying machine made out of a brake cylinder).

Still, the geek siren song of the Bay Area beckoned. In 2006 she pitched a talk to CodeCon, the same San Francisco–based anticonference for hackers where Kay Aull a few years later described her DIY gene test. Patterson was accepted. After her talk her boyfriend proposed to her during the onstage Q&A. She said yes.

Her now husband, Len Sassaman, lived in the Bay Area, and she stayed. Sassaman has serious hacker credentials himself. He cofounded CodeCon with Bram Cohen, who is best known for inventing BitTorrent, the popular peer-to-peer file-sharing protocol and scourge of the music-, movie-, and television-industrial complex. He has a long affiliation with the cypherpunks, a loose network of computer geeks focused on improving online security and privacy through cryptography. The group has always enjoyed a relationship of mutual distrust with authorities, who view cryptography they cannot crack as a national security risk, not just a tool for pirating copyrighted work. (The cypherpunks count hacker and Wikileaks founder Julian Assange as one of their own.) Like DIYbio in the present, an online mailing list formed the hub of the cypherpunk movement in its 1990s heyday.

The cypherpunks' antiestablishment credo was summed up in 1993 by activist Eric Hughes in "A Cypherpunk's Manifesto," which lays out the terms of privacy in a digital age and the need for individuals to claim that right for themselves through their own ingenuity and initiative as software developers. Patterson wrote the biopunk manifesto as a creative remixing of "A Cypherpunk's Manifesto." Hughes wrote: "We must defend our own privacy if we expect to have any." Patterson says: "We have questions, and we don't see the point in waiting around for someone else to answer them."

The day after Patterson spoke, the biohackers held a fair, or faire in the preferred geek parlance. At least one project was deeply serious: Johns Hopkins–trained geneticist and physician Hugh Rienhoff described his ongoing efforts in his home lab to diagnose his seven-year-old daughter's unidentified genetic disorder. Other exhibits were done in a spirit of play. Of course, it takes a biohacker's peculiar sense of fun to teach middle schoolers how to use glowing bacteria to decorate little metal buttons to pin to their backpacks.

At the registration table, undergraduates made a soupy mess using dish soap, rubbing alcohol, salt, and a lot of squashed fruit to extract skinny strings of DNA from strawberries—the quintessential parlor trick of DIY biotech 101. The strawberry genome is much shorter than the human—about 200 million letters compared to 3 billion. But unlike humans, who only have two copies of each chromosome per cell, strawberries have eight. In other words, strawberries are stuffed with DNA. They are also easy to smoosh, which makes them perfect for table-top experiments that even elementary school kids can do. The strawberry DNA demo has an elegant simplicity that serves as a goopy gateway into genetics. Using a few household chemicals, anyone can break open a strawberry cell's walls and isolate its genetic core. At the end, you twirl the stringy stuff of life on the end of a chopstick. Suddenly DNA ceases to be an abstract concept, a sci-fi–tinged computer-animated demo of a double helix. Here it is, right in front of you: the physical crux of everything that lives.

Patterson had her own table, where she walked purple-gloved volunteers through her signature experiment. Using only simple household materials (almost), she proposed to extract plasmids from the bacteria used to culture yogurt. Plasmids are typically loops of DNA that exist in bacteria separately from the cell's chromosomes, the DNA required by the cell to replicate itself. Plasmids can replicate on their own independent of a parent, and bacteria do not need them to survive and perpetuate. But bacteria do exchange plasmids with one another, which makes them the ideal tool for genetic engineers who want to insert genes from one kind of cell into another. Patterson calls plasmids the workhorses of genetic engineering. Since genes were first spliced in the 1970s, plasmids have been seen as the most useful tool for transferring snippets of DNA from the cells of one organism to another. The bacteria whose plasmids were used in the first gene-splicing experiments were *E. coli*, which continue to be the go-to bacteria for biotechnology.

But *E. coli* is a little fragile for kitchen lab use, despite its lethal reputation. This is why Patterson likes to work with bacteria best known for their use in the kitchen. People have slurped *Lactobacillus* bacteria unknowingly for thousands of years in the strangely edible form of spoiled milk we know as yogurt. *Lactobacillus* are a little tougher to work with than *E. coli* but also more durable. Patterson likes to point out that they also have the advantage of being readily available. Biopunks never tire of observing that the raw materials of biotechnology are always just a supermarket away.

Patterson used a micropipette to siphon off a little of the liquid sloshing around the top of her container of Brown Cow peach yogurt and inoculated a small test tube of skim milk with the contents. This pretty much amounts to what it sounds like: injecting the bacteria-rich liquid into the milk that has sugar the bacteria will feed on and then multiply. As Patterson also likes to point out, this is the first step you would take if you wanted to make homemade yogurt. But the glass and gear spread out across the table showed she had something very different in mind.

Next to the skim milk was a bottle of Everclear brand alcohol, salt, paper towels, and multipurpose contact lens solution. There was a netbook computer, measuring cups and spoons, plastic wrap, the yogurt, nine small beakers, a miniature centrifuge, an open-source gel electrophoresis box, and a digital scale that everyone agreed would be easiest to buy at a store that also sold bongs and rolling papers. Under the table was a small cooler. A few pairs of purple gloves were on the table but "we're kind of low on safety equipment, which I'm concerned about," Patterson said. She also did not have what she needed to maintain the kind of sterile environment that even *Lactobacillus* need to survive the rigors of tabletop experimentation. To fix that situation, she instructed a volunteer to go down the hall to an FBI recruiting booth, where special agents were handing out souvenir hand sanitizer to lure biogeeks into their biodefense program. The pen-shaped vials read: "Today's FBI—It's for you."

As Patterson worked, a few veteran scientists who have worked in corporate biotech labs and held tenured university positions stood off to the side and watched. They smiled the way a parent does watching a child at play. The looks were not condescending. These were men who looked like they were remembering their first day in a lab, or that rainy afternoon decades ago when they first scattered chemistry sets across their own kitchen tables. They were remembering that first raw curiosity that had brought them all this way.

Which of these goofs is going to lead to the next great discovery? Will these playful projects pile up until they reach a mystical critical mass when some alchemy of innovation transforms fun into serious and important work? In Patterson's ideal world, the fruits of creativity will sprout from below rather than fall from above. Moreover, if innovation emerges from the ground up, then perhaps its benefits will come more quickly to those at the bottom.

CHAPTER 5

Field Testing

On Venezuela's high plains, poor farmers scratch out a living from tough tropical soils. Cowboys herd cattle across endless grasslands battered by a yearly cycle of flood and drought that supports a menagerie of exotic wildlife but seldom a decent livelihood. Rodents survive especially well in this environment of extremes, among them the four-foot-long flat-muzzled capybara, the largest rodent in the world.

But in Guanarito, a small farming town in the west central part of the country, more familiar rodents are the ones that matter. In 1989, doctors began encountering farmworkers from here who were suffering from relentless fevers. They were weak and dehydrated. They bled from the nose and gums. They vomited blood. A third of them died. Many of the sufferers were first diagnosed with dengue fever, a common tropical disease around the world, including tens of thousands of cases diagnosed each year in Venezuela.

Eventually lab tests would identify a new culprit researchers called Guanarito hemorrhagic fever, after the town where the infection first appeared and where it first victimized a poor rural population ill equipped to deal with the onslaught of a dangerous new infection. Scientists working in the Venezuelan hinterlands discovered the virus that caused the disease in the feces of mice and rats, which farmworkers encountered most often while clearing the fields during the dry season, from December to March, the months when more than half the reported annual cases would appear.

Guido Núñez-Mujica was a small boy when the first cases of the infection appeared. He grew up and still lives in Mérida, an urban outpost of a quarter million people in the Venezuelan Andes about 130 miles from where the infection originated. He came of age under the ascendance of Hugo Chavez and enrolled as an undergraduate at the Universidad de los Andes (University of the Andes), a site that in recent years has seen clashes between pro- and anti-Chavez students turn violent. Núñez-Mujica has no special love for his country's leader, yet he and the president agree on at least one thing: The rural poor in their country have suffered because they lack access to the basic benefits of modern life that the more affluent take for granted. Chavez believes the state can right this inequality. Núñez-Mujica believes in biotechnology.

Now twenty-six, Núñez-Mujica specializes in applying computer science to biological problems. He describes himself as "a nonconformist, forward-looking geek." His stake in making the tools and techniques of biotech accessible to more people is more personal than that of most of his peers in the United States and Europe. Biohackers in the developed world are typically outsiders by choice. Companies and schools have the tools, but do-it-yourself biologists are willing to trade that access for the freedom to do what they want the way they want to do it. At his university, Núñez-Mujica says labs have one-tenth the budget of a comparable school in the United States. At the same time, thanks to monopolies on the distribution of lab equipment, gear costs twice as much, he said. Ambitious Venezuelan scientists faced with these conditions often flee, creating a chronic brain drain. Núñez-Mujica made a different choice. Instead of moving to labs in other countries with ten times the budget, he wondered what would happen if he could build gear that cost a tenth of the price?

Núñez-Mujica was inspired by the desire to invent more options for those like himself with access to less. He was also inspired by his lab's inventiveness in the face of scarcity.

In Central and South America, Chagas disease is endemic. Insects known as kissing bugs carry the single-celled parasite that causes Chagas. The insects got their name from their tendency to bite people on the lips while they sleep. Chagas causes only mild symptoms in its acute phase, during the first few weeks or months after someone is bitten. Afterward, the pathogen can lay dormant for years, even decades. Eventually, however, untreated Chagas enters the chronic phase. In the worst cases the colon and heart can become inflamed, leading to congestive heart failure and sudden cardiac arrest.

Because patients must start taking drugs to treat Chagas early to rid themselves of the infection, prompt diagnosis is crucial. At the University of the Andes, the lab Núñez-Mujica worked in as an undergraduate developed a kit. When the parasite infects a victim, the body tries to fight it off. The lab's kit contains all the chemicals and tools needed to test for the antibody the human immune system creates specifically to battle the Chagas invader. To keep costs down, the lab managed to have all the chemicals made locally. If they could pull off such a project with the limited resources they had, Núñez-Mujica wondered why they couldn't do more.

To test for the antibody, the lab's kit uses a standard technique known as an ELISA (enzyme-linked immunosorbent assay). The ELISA works by filling tiny wells in a plastic tray with a protein that acts as the marker for the parasite; a sample of the patient's blood is added. If the patient is infected, the antibodies in the blood attempting to fight the disease will attach themselves to the parasite's protein, and the test will signal a positive result.

ELISAs are reliable, but they are not versatile. Each kind of ELISA must use the specific protein for the disease a doctor wants to identify. Testing for a variety of diseases requires multiple tools and multiple tests. What's more, corralling the proteins needed for each test takes money and work.

Núñez-Mujica envisions a different kind of diagnostic, at once more sophisticated, simpler, and easy to take anywhere. His test would seek the genetic signature of the parasite itself. The direct approach would seem the most obvious. But lugging the necessary gear into the jungle would be difficult—unless, that is, your crucial piece of equipment was the size of an iPhone. For Núñez-Mujica, this insight into the need for portability and simplicity would anchor his grand plan for an off-the-grid diagnostic rapid response kit that he hoped would make the tools and techniques of biotech truly available to everyone.

Joseph Jackson could be an investment banker. He could be a Senate staffer. He could be vice president for business development at a stable, midcap biotech firm with several phase one and two clinical trials in the works and several promising experimental drugs in the pipeline. Instead, he is trying to sell a small metal box the size of a few iPhones stacked one on top of the other, for as little as possible, to people with no money.

Jackson graduated from Harvard in 2004, where he studied economics and politics. Unlike his classmates, he did not head into banking, consulting, or government. He was too restless to follow the typical path. And he had experienced a moment in recent history he couldn't get out of his mind.

When Jackson started college, students across campus and across the country were downloading music via Napster. Shawn Fanning, a student at Northeastern University just across the Charles River, had launched his music file-sharing service in early 1999, while still a teenager. Jackson's friends were busy filling their hard drives with free music, but he had a feeling that something more important was happening.

More than just undercutting the record label middle man, Jackson was struck by the way Napster let any two people with a similar interest find each other, out of a pool of millions of users. In the case of Napster, users were coming together over a song. But Jackson saw that the song was just one form of content, one type of idea, one category of information that this new way of networking could make available to mutually interested minds. Peer-to-peer networking wasn't just a way to share music, he decided. It was a way to share knowledge of all kinds.

These instincts coalesced into a single-minded sense of purpose after he heard a talk by Yochai Benkler, at the time a Yale law professor who was an early champion of the so-called information commons. Benkler wrote that the greatest drag on progress of any kind comes from locking knowledge away behind the castle walls of intellectual property law. In his influential 2006 book, *The Wealth of Networks,* Benkler argues that the kind of open-source collaboration made possible by the Internet—he calls it "commons-based peer production"—is not only an inevitable consequence of twenty-first-century interconnectedness but also the best way to make money. In one oft-cited example, Benkler observes that IBM in 2003 made $2 billion helping customers run Linux, the open-source operating system that anyone can download for free. That same year, the company made less than half that from intellectual property licensing and royalties, despite generating more patents every year for the previous decade than any other company in the country. Jackson saw no reason that the same approach could not apply to biotech.

Jackson began exploring these ideas more deeply at the London School of Economics, where he studied philosophy and the history of science. In the process, he began delving more deeply into transhumanism's radical technological optimism. At a transhumanist conference in 2005 in Caracas, Jackson first met Núñez-Mujica; they met again at a similar conference in Chicago two years later. At the time of the second meeting, Jackson was starting to try to figure out how to transform open-source science from an intellectual ideal into a working movement. Núñez-Mujica wanted to apply open-source collaboration to drug development. They watched each other's presentations and recognized they shared the same hyperarticulate passion. Transhumanists dream big about the future—centuries-long life spans, an end to world hunger, human brains with the computing power of a microchip. And they do not believe in leaving these grand achievements to some distant future generation. They believe their goals are not science-fiction dreams, just engineering problems. It was in this atmosphere that they decided to chase Núñez-Mujica's ambition to create inexpensive diagnostics for the developing world, in the process creating a tool Jackson hoped would wedge open the door another inch to peer-to-peer biotech.

The LavaAmp prototype is made from sheet metal. It is a box that has a small cylinder jutting out at one end encircled by two small wires. You can power it with AA batteries or by plugging it into the USB port on your laptop. It is divided into three chambers. The basic task the box performs does not sound especially remarkable: It heats up, then cools down, then heats up again. Liquid goes in, liquid comes out, and it looks pretty much the same. A small child might mistake it for a toy at first. But after a few seconds of poking and shaking, they would probably get bored and move on.

In short, the LavaAmp seems mundane. Yet this lack of surface sex appeal is exactly what makes it so remarkable. Jackson and others have crafted a machine that testifies to the unbelievable complexities unraveled by life scientists over the last three thousand years.

The LavaAmp is perhaps the world's smallest version of what is known as a thermal cycler: a basic piece of biotech equipment that rapidly replicates DNA. The regular rise and fall in temperature of a thermal cycler makes possible what's known as a polymerase chain reaction, or PCR. The discovery of this reaction and how to manipulate it, after its invention in 1983, was one of the fundamental achievements that made genetic engineering on an industrial scale truly possible. The power of PCR is not just that it duplicates DNA, but that it allows researchers to copy specific snippets of DNA. Thermal cyclers let biotechnicians who know the right combinations of chemicals whip up vats of just the gene they want to splice. Biotech drugs, genetically modified crops, and synthetic biofuels all use PCR as a basic part of the manufacturing process.

Historically, they have also cost labs tens of thousands of dollars. And in some ways understandably so. They may sound like little more than advanced Crock-Pots, but PCR requires precision to work well. Even as they become commonplace, what thermal cyclers do is still pretty incredible.

Since first congealing in the primordial ooze billions of years ago, DNA has spent all but an infinitesimal slice of the earth's history reproducing itself on its own schedule and in its own combinations. Now we have machines that let us order up the DNA sequences we want, when we want them. A machine like that would have to cost a lot, right? After all, even God has one in His toolbox.

Jackson says no. And the LavaAmp is his pièce de résistance. When he's pitching the device, he can be, as Seinfeld would say, a "close talker," leaning in the way a politician might to suggest empathy and engagement. His voice carries a hint of his native South Carolina, and he is quick with his talking points, a skill he credits to the whirlwind of meetings and pitches that are the day to day of any young entrepreneur. One fact he doesn't often mention is that as a kid in Columbia, he met Kary Mullis, a family friend and Carolina native who had just been awarded the Nobel Prize in chemistry for inventing PCR. Perhaps Mullis's 1993 Nobel lecture planted a seed in Jackson's subconscious as a boy that later flowered into his current attitude toward science. In the lecture, Mullis recalled how he spent his days at his South Carolina high school in the early 1960s: "We were allowed free, unsupervised access to the chemistry lab. We spent many an afternoon there tinkering. No one got hurt and no lawsuits resulted. They wouldn't let us in there now. Today, we would be thought of as a menace to society."

The LavaAmp is expected to retail for less than $100. The goal: Let scientists manipulate DNA just as easily in a Congolese village as a Bay Area wet lab. Or at least about as easily. A basic tenet of do-it-yourself biotech is that close enough is good enough.

Jackson and Núñez-Mujica believe the LavaAmp's portability, versatility, and price will ultimately make it a key component of a developing world diagnostics kit. Unlike the typical kits used by public health workers today, tests for different disease markers using the LavaAmp kit would not require a separate arsenal of chemicals for each germ. Instead, a doctor or a public health tech would only need a small set of primers, short strands of the genetic alphabet that mark the start- and endpoints of the DNA snippet to be clipped and replicated by a thermal cycler like the LavaAmp.

In Núñez-Mujica's ideal vision, a health worker could visit a place like Guanarito as soon as an infectious outbreak occurs. Armed only with the LavaAmp, a DNA-reading chip, and a few vials containing primers for hemorrhagic fever, Chagas, and dengue, the worker could quickly test for all three disease-causing pathogens and get an answer within minutes. The materials are cheap, the gear portable, and the techniques efficient. An impoverished country's government could afford to buy the kits by the gross. And an entrepreneur would not need millions of dollars in venture capital that would in turn force him or her to make tens of millions of dollars to provide a return on that investment. This is important when you're making a product that will be marketed for the poorest people in the world. The other key component would be a small chip that could read the DNA samples frothed up by the LavaAmp. The pieces would all be small enough to wear clipped to a belt.

This ragged around the edges approach to biotech is the source of much of the skepticism directed at outsider biology. If you trick out a Honda Civic *The Fast and the Furious*–style in your home garage and it does not start, grab a socket wrench and start cranking until it does. In the meantime, your car won't die, at least not literally. Can't get your awesome new iPhone app to stop sending your credit card number out to all your Twitter followers? Don't sleep until you've debugged. You can't kill your code, not in a shuffle off this mortal coil kind of way. Only in biology can your project truly perish. As a consequence, keeping your gear tightly organized and sterilized in a clean, well-lit place is traditionally viewed as an unshakable premise of wet lab work.

In reality, spotless labs bathed in bright white light over shiny black benches are more the stuff of Hollywood sets and public relations brochures even at some of the nation's top research universities. In impoverished countries, where electricity and running water cannot be taken for granted, sterile lab environments are often an unaffordable luxury, especially in the field, where the LavaAmp's developers believe it will do the most good.

In early 2010, the LavaAmp's inventors decided to crowdsource the funding they needed to nudge the device out of the prototype stage and toward the mass market. They turned to the Unreasonable Institute, a Boulder, Colorado-based incubator for social entrepreneurs. The institute's Web site lists its expectations for projects it backs: "All ventures must effectively address a social or environmental issue, be financially self-sustaining within 1 year, have a model which can be scaled out of the country of origin within 3 years, and must eventually meet the needs of at least 1 million people."

Núñez-Mujica looked at the daunting criteria and said, "That's us." To attend the Unreasonable Institute, entrepreneurs compete to be among the first twenty-five who can raise $6,500. Each gets a page on the Unreasonable Marketplace Web site to promote his or her vision. On the LavaAmp page, Núñez-Mujica described the device as a "rugged, accessible, DNA diagnostic tool to quickly respond to pandemics and neglected diseases."

"Diagnostics for the developing world" has become a rallying point for outsider biologists. Some on the scene have always tried to stress that ease of access to the tools and techniques of biotech is just half the goal. The other half is to make the output of all that innovative effort accessible to as many people as possible. And making biotech accessible means making biotech affordable.

Biotech drugs are famously the most expensive pharmaceuticals on the market. A year's supply of Genentech's top-selling Avastin anticancer drug can cost nearly $100,000. Genzyme of Cambridge, Massachusetts, grew from a small start-up into one of the country's five largest biotechnology companies by charging $200,000 a year for drugs that treat extremely rare diseases. The push by biohackers to cut the costs of biotech subverts the traditional venture capital–driven business model. But it is also about being practical.

Suffering creates the marketplace for the pharmaceutical industry. Drug developers make money by developing products that alleviate suffering. Yet some health problems lack a financial incentive to solve them, such as the millions in Africa who go without immunizations because they cannot afford vaccines. As a result, the Bill and Melinda Gates Foundation has stepped in with massive philanthropic contributions for mass vaccinations on the continent, effectively setting nations' public health agendas from the outside.

There would seem to be similarly little profit in making diagnostic tools for rare diseases or even common ailments that afflict only the developing world, whose inhabitants cannot pay for the devices. But Jackson and Núñez-Mujica propose a radically different strategy than Gates to solve a similar problem.

The pair first could try to join the upper strata of the rich in the global economy by following the Gatesian arc of traditional entrepreneurship. Then they could pour that money into whatever social cause they wanted. Instead, they want to make profit irrelevant to philanthropy. They want intellectual capital to make profit beside the point. They understand it doesn't do much good to teach a man to fish if he can't afford a net. But give doctors in sub-Saharan villages a nearly free PCR machine and the know-how to use it, and they no longer need to rely in the same way on the developed world's pharmaceutical-industrial complex. They will have become networked outposts in a newly decentralized public health infrastructure, where the barrier to entry for biotech has shrunk to the size of a LavaAmp.

Jackson and Núñez-Mujica are dreaming big, and they know it. Just getting the device to work is challenge enough; even then they face real barriers to seeing the LavaAmp widely adopted in the field. The price of proving its medical validity, of getting a diagnostic test approved for use, could cost millions. But Núñez-Mujica does not see the issue as optional.

"It's not like the next iPod or next iPhone. If they don't happen, so what? These technologies are good enough," Núñez-Mujica says. On the other hand, he believes that if biotech does not make radical leaps in innovation and accessibility, "this means millions of people will get sick and die."

This sense of urgency keeps the pair committed. "So as long as we can hold out," Jackson says, "we're going to do it."

CHAPTER 6

Cheap Is Life

Mac Cowell has said that inexpensive tools like the LavaAmp are valuable because they will help build what he calls biotech intuition, which he believes will ultimately lead to innovation.

Meredith Patterson writes in "A Biopunk Manifesto": "A thirteen-year-old kid in South Central Los Angeles has just as much of a right to investigate the world as does a university professor. If thermal cyclers are too expensive to give one to every interested person, then we'll design cheaper ones and teach people how to build them."

In DIY subcultures of all kinds, hacking the cost of materials, equipment, and labor is always a top priority. Money is seen as a barrier to creativity, the inability to afford the right tools an artificial constraint on ingenuity. For a long time the price of doing biotech far exceeded anything a hobbyist could hope to afford. More recently, eBay and Craigslist have made it easier for DIYers to get steep discounts on used gear, which biohackers say has been especially plentiful as companies sink in the wake of the recession. Still, even preexisting equipment cleverly sourced on the cheap can feel like a bit of a cop-out among the most committed gearheads.

Fortunately for them, the rise of do-it-yourself biotech has coincided with the growth of open-access manufacturing. Community-based machine shops like TechShop in Silicon Valley allow would-be inventors pay a subscription fee to use everything from lathes and milling machines to a computer-controlled vinyl cutter and an industrial sewing machine. Three-dimensional printers take devices from computer desktop to prototype in one step. The Internet enables inventors to track down the best deals on fabricators, send their blueprints anywhere in the world, and get a prototype back via FedEx. Open-hardware geeks are working on ways to take that process a step further by combining digital specs with 3D printers to make three-dimensional, fully assembled, touchable, holdable objects that are as downloadable as music or movies. All of these efforts ultimately converge toward a single point: To put tools into the hands of as many people as possible—as many who care enough to want it in the first place.

Tito Jankowski grew up on the Big Island in Hawaii. His parents are painters. He is well over six feet tall and has a surfer's shock of shoulder-length blond hair. And he is psyched. Of all the biohackers I've met, Jankowski is the most eager to push the idea that biotech can be fun. He does not necessarily need to change the world through genetic engineering, though he would not have a problem with that. But he does want to change the way people think about and interact with biotech. And his strategy is revolution by enthusiasm.

Jankowski graduated from Brown in 2008. He studied bioengineering and helped start Brown's iGEM team. He doesn't exactly count as a biotech amateur. Yet when he got out of college, he took a programming job with a giant management consulting firm and moved to Sacramento. Rather than trying to get paid for what he loves to do—often a quick path to killing the joy that drew you in the first place—Jankowski opted to keep it fun. The way to keep the love of biotech strong was to keep it a hobby.

Of course, the very idea that biotech could be a hobby was even stranger in 2008 than it is today. But Jankowski in his blithe way never worried that there was anything weird about it. Instead, he decided to bring together open hardware and DIYbio to make access to biotech tools easier.

In a bare, cold garage in Sacramento, Jankowski and his friend

Norm Wang explained to me that biology is hard. Even a simple process like running a gel to analyze the size of DNA fragments, as in Kay Aull's hemochromatosis gene test, is easy to screw up. If any one step in the process goes wrong, you have to start over again. And using the gear found in most labs, you will not likely know you made a mistake until your end result fails to appear.

"It's sort of like driving for an hour [with your eyes closed] and then opening your eyes and saying, Where am I?" Wang said.

To get around this problem, the pair have built what they consider a better gel box. They use a grid of LED lights instead of ultraviolet ones and a dye that enables them to track their DNA samples as an electric current passing through the clear gel pulls smaller fragments farther toward one end while leaving larger, heavier fragments behind. They will know sooner rather than later if they have to chuck it all and start again.

Jankowski and Wang did not invent this technique. They are not the first to build a box that lets them keep better track of their samples. They could have spent about $1,200 to buy one from one of the big corporate biotech supply conglomerates that does the same thing, they say. But as two guys in their twenties not long out of college, that kind of money is just not lying around. Instead, they did what a lot of guys in Northern California in their twenties do: They started a company.

Pearl Biotech makes easy-to-track gel boxes that sell for $200 each. That is, if you want Tito and Norm to make them for you. The price is great compared to the going rate from big suppliers. But you can have the same box for even less. A lot less. Because also like a lot of young entrepreneurs in Northern California, they are happy, even eager, to give you something free—in this case, the blueprints for their box.

The open-source model of innovation has held sway over software development for so long that it rivals the more traditional closed approach as software-engineering orthodoxy. The movement first drew mainstream attention in the 1990s in the form of an upstart answer to Windows bloat and boredom. Programmer volunteers from around the world built Linux, an operating system that its army of creators believed ran PCs better than the OS that had made the richest man in the world his fortune. When working for free, the argument went, the only motive is innovation, the pure intellectual pleasure of building a tool that works just like you want it to, paid for in sweat equity and perhaps the prestige bestowed by your fellow geeks if you pull off an especially cool hack.

In the first decade of the twenty-first century, the Firefox Web browser was the most celebrated and widely used open-source achievement. Linux even now requires a steep learning curve that limits prospective users to code warriors and copyright reformists who cannot abide the idea of a corporate-controlled computing experience. Firefox was different. Anyone could figure it out. It was free. It was easy to customize. And it made surfing the Web better. As a result, Firefox showed that open-source could have true consumer clout. Over the past five years, Internet Explorer's market share has eroded, despite coming bundled with Windows, the world's most widely used operating system. Companies that wanted to stay in the innovation game had to take notice. People want better products, and open-source seemed the way to build them.

In 2008, Google did not surprise anyone when it released the code to its own new browser at the same time it released the browser itself. Google also set the standard for collaborative crowdsourced innovation on a more modest but more accessible scale by opening the lid on many of its products with application programming interfaces, or APIs. Countless Google map mashups were just one species in an ecosystem of creativity made possible by giving tinkerers an easy way to put their own spin on the company's products, and they built wildly original new tools in the process. Twitter took APIs to the next level, opening up its one and only product from the start. That decision has led developers to create hundreds of custom ways to use and experience Twitter. Its executives have reported that their API traffic is double their Web traffic—all through services designed by others to whom Twitter did not pay a dime.

The success of open-source software has generated much excitement about applying its methods to innovation in other industries. But open-source hardware has not kept pace. Jankowski and Wang believe inventors of actual physical things, unlike coders, have simply not figured out how to talk to each other.

"We don't have a lot of the language that we have for software," Jankowski told me as we sat around a table littered with their gel box's few electronic parts. "It feels like those early days, when it's still hard to do stuff," Jankowski said.

The pair, both trained as engineers in college, had to learn basic manufacturing processes from scratch to turn their blueprints into boxes. Unlike code, you cannot both design and fabricate a three-dimensional object on a laptop at a café table ("yet," you can hear the eternally optimistic chorus of technophiles chime in).

Instead, Jankowski had to dig into the workaday world of real physical stuff. He figured out that he could have the plastic cut by a sign shop down the street. Meanwhile, Wang began fiddling with an Arduino, an open-source electronics prototyping kit—a kind of Lego for electronic circuits—prized by hackers of every persuasion. It took him a day to figure out a simple controller for the LED grid that lights up the gel and stimulates the dye staining the DNA in the box. It took him at least as long to figure out how to get his design to a manufacturer in China, the cheapest way to mass-produce the boards. Once they had the parts, they began building the boxes in small batches by hand.

The process is arduous and the revenues slight. Neither is making a living off Pearl Biotech, perhaps in part because they charge about $1,000 less for their box than the price of the standard name-brand variety—even though they piece each one together by hand. They shun mass assembly because pricey molds force you to stick to a particular design, when the whole point is to make the box tweakable. Wang is getting a PhD in bioengineering from the University of Hawaii in Honolulu, and he cannot stand that the tools he uses every day in the lab come weighed down with proprietary licenses from their manufacturers that prevent researchers like himself from modifying their gear. But like so many others passionate about open-source, money is not the point, at least not yet.

"We're building a philosophy, rather than just making something cheaper and cheaper," Wang said.

Jankowski's next project was even more ambitious and attracted more attention as a result. Gel boxes are useful and necessary in any biology lab. But they are tools for analysis. They take DNA apart. They do not perform the fundamental revolutionary trick of biotech: putting genes back together again in ways nature never figured out on its own. Making gene splicing as well as gene splitting possible in the hobbyist's lab would be the next step. And the first tool any biotech lab needs to make that happen is a photocopier.

In the earliest days of gene splicing, performing PCR meant hours of moving test tubes back and forth between baths of water at very specific temperatures. It was an imprecise way to accomplish a task that required great precision. The first automated PCR machines did for labs what Xerox machines did for offices. They were such a welcome convenience that those who made them—and controlled their patents—could charge a premium that labs would gladly pay to alleviate the tedium and high failure rate for their experiments. Even after PCR became routine, that premium never really went away. The machines became more advanced, more precise, more digital. But they still cost a lot. Enter OpenPCR.

The OpenPCR machine's outer case is made of wood, the most obvious difference from standard mass-produced plastic versions of the device. The size of a toaster, it consists of sixteen wells for the vials of DNA to be copied. A digital display tracks the heating and cooling cycles, and the machine will tweet or text you when it's done. Jankowski and OpenPCR partner Josh Perfetto designed and built the guts of the machine using Arduino. Since the greatest expense of PCR is the chemicals involved and not the machine itself, they are also developing software designed to optimize reactions to minimize both mistakes and the volume of chemicals needed.

Whether OpenPCR will ever let Jankowski quit his day job is difficult to predict. Customers can buy kits and build the machines themselves, or for more money the pair will put it together for you. The most they plan to charge is $400, compared to the several thousand dollars needed for the typical machine bought from a big supplier. Because open-source is the whole point of OpenPCR, anyone who wants a PCR machine can download the blueprints and build one without paying Jankowski or Perfetto anything. Jankowski said someone industrious enough could take the plans, hire workers, and begin mass-producing OpenPCR machines as a for-profit business. Inspired by open-source standard-bearers like Linux and the open-hardware movement, Jankowski believes that making biotech available to anyone beats getting rich. He has not invented a new technology in hopes of turning a profit; instead, he is trying to use an old technology in an innovative way. And he's banking on convincing a critical mass of fellow idealists to join him.

In June 2010, he got off to a good start. Among DIY enthusiasts and indie creatives in nearly every medium, Kickstarter.com has become a go-to site for funding projects. The concept is simple: Someone pitches a project, whether a movie, a pie-baking contest, or a DNA-copying machine. The project's creators say how much money they need to raise and set a one-to-ninety-day deadline for raising the amount. After that they try to draw attention and support any way they can. Anyone who comes to the site and finds the project interesting can give however much they want. The catch: If the creators don't raise the full amount by the deadline they set, they don't get any of the money pledged at all. According to Kickstarter, the all or nothing concept lowers the risk for everyone: "If you need $5,000, it's tough having $2,000 and a bunch of people expecting you to complete a $5,000 project."

Meredith Patterson spread the word on her blog. Jankowski announced the project on the DIYbio mailing list. Pledges began to trickle in. But Jankowski knew that the key to support lay in enlisting the larger DIY community. He sent out e-mails to everyone with any pull among makers, and scored when Dale Dougherty, the creator of the Maker Faire and founder of *Make* magazine, posted a link to OpenPCR's Kickstarter page on Twitter when they were about halfway to their $6,000 goal. Dougherty had about 1,900 followers at the time, a respectable number. A few hours later, tech publishing superstar Tim O'Reilly retweeted the link to his more than 1.4 million followers. Within two days, the pair had topped $6,000—more than a month ahead of their July deadline. By the end of the fund-raising drive, the pair had received more than $12,000 in pledges—twice their original goal. Jankowski said the extra funds meant they would likely be able to take the machine to the next level of technological sophistication, known as quantitative polymerase chain reaction, or qPCR, in which the machine detects and quantifies the amount of target DNA being produced as the reaction goes forward.

Still, however good or cheap OpenPCR becomes, doing biotech takes more than just lifeless tools. Splicing genes or even just making copies requires chemicals to cut and copy the DNA—the sometimes gooey, sometimes toxic, and generally expensive wet stuff that makes biotech different and more difficult than other kinds of engineering. Once up on Kickstarter, the pair ran into skepticism right away from one commenter, who pointed out that in PCR, the machine is not the main expense. In the same way that cell phones don't do much without minutes, PCR machines need the right chemical reagents to become anything more than overpriced Crock-Pots.

Solving the reagent problem has become Jankowski's next big task for moving open-source biotech forward. But he believes he has at least one hack up his sleeve that requires little cash. As he learned in his own garage, experiments often do not go right the first time. Or the second time. Each failed experiment is money spent. And these failures do not apply only to untested approaches to new science. Often the routine steps needed to get from an idea to trying something new do not go right. Contamination is the scourge of every wet lab: Fail to follow all the steps exactly and some unwanted microbe will get in and kill the cells you want so much to splice. Jankowski believes that short of inventing more open-source machines to automate these intermediate steps, the key to fail-safe, and hence cheaper, experiments is open protocols. In other words, let everyone take a whack at the instructions until the wisdom of the crowd gets them just right. And then publish them in as accessible a format as possible, like a cookbook. Or a comic book.

Though gene splicing through direct manipulation of DNA was invented nearly forty years ago, genetic engineering as a concept still has a futuristic feel. The unraveling and rearranging of life's genetic guts does not seem like a part of your everyday experience. Or at least not that you realize. Of course, genetically modified crops are ubiquitous on the American landscape. More than half the acres of corn and cotton and more than 90 percent of the acres of soy planted in the United States in 2009 were sown with genetically modified seeds. Biotech drugs have become a standard if expensive part of the pharmacopoeia. As biofuels become more commonplace, genetically engineered fuels could fill our cars. For a technology that seems so alien, recombinant DNA is hardly obscure.

Jankowski hopes that tools like his will help more people overcome the feeling of disconnection they have from their DNA. The more he can get gel boxes and thermal cyclers into the hands of those with little experience in molecular biology, the more minds he hopes will recognize and contribute to what he sees as biotechnology's great promise. He especially hopes to introduce young people to the tools and techniques of biotech in a way that makes gene tweaking as much a part of everyday technology as texting.

"In past years maybe it did make sense to pay $5,000 for a PCR machine. But I think the context is changing," Jankowski told me. "You're getting people who are interested in contributing to biology but don't necessarily have the typical setup. People are either working in garages or community labs or high schools. That's the new context. And that's why this stuff makes sense now."

CHAPTER 7

Homegrown

For most of the history of the life sciences, major discoveries were made in labs less sophisticated than today's kitchens. And not kitchens stocked by biohackers with secondhand lab equipment, but kitchens simply intended to prepare food. Laser meat thermometers. Handheld blenders. Microwave ovens. Ziploc bags. A steady supply of electricity, natural gas, clean water, and refrigeration. Researchers developed vaccines, the germ theory of disease, and the foundations of modern genetics without any of these conveniences. What average Americans would see now as primitive living conditions did not prevent the discovery of most of the basic principles that still guide biology.

Yet one technological advance stands above all others as having done the most with the least. Forget running water. What about the biohackers who changed history—in a sense, made history possible at all? The invention of agriculture still stands as the world's premier biohack, the ultimate example of amateur innovation.

Why humans began planting and harvesting crops about ten thousand years ago after more than two and a half million years of hunting and gathering remains unknown. Competing hypotheses abound, many related to a changing climate following the end of the last ice age. Others point to evolution as a driving force, in which plants, animals, and humans found settling down to be mutually beneficial for all. Other theories point to tribal politics, in which the ability to hoard food became the key to power. Regardless, no one person or group "invented" agriculture. This was one innovation that was clearly crowdsourced.

Whatever the reason, many *Homo sapiens* ceased their migrations and settled down to grow crops and tend domesticated animals. Since then, barnyard tinkerers have spent millennia fiddling with nature to create bigger, stronger, tastier, more plentiful crops and livestock. That dynamic changed dramatically with the introduction of genetically modified crops in the 1990s. Any kind of crossbreeding in agriculture has always meant the mixing of genes into new combinations. Only in the twentieth century did scientists begin to truly understand the underlying biochemistry, and only in the late twentieth century did scientists gain the ability to alter those genes through a direct manipulation of DNA. With that ability came a new business model of agriculture, under which genetically modified seeds were the intellectual property of the companies that made and sold them, no different than computer software or a new design for a hybrid car engine. To use that property farmers must pay the companies licensing fees. In exchange, the farmers gain the ability to grow crops with resistance to pests and herbicides built right in their DNA.

Even as U.S. farmers began to blanket their fields with genetically modified crops, this widespread release of biotechnology into the world came under heavy criticism. A fearful public fretted about so-called Frankenfoods, supposed perversions of nature that would spread and take over. Environmentalists feared the new varieties of biotech corn, soy, and cotton could invade ecosystems and threaten biodiversity. Organic farmers worried that so-called transgenic crops could contaminate their own fields. Anticorporate activists claimed that turning seeds into intellectual property rather than holding them in common in keeping with millennia of tradition allowed a few companies—especially agribusiness giant Monsanto—to monopolize agriculture. The public worried that genetically modified foods could be dangerous to their health. The backlash against these new bioengineered breeds was especially strong in some poor and developing nations, where many viewed the technology as just the latest form of Western corporate imperialism. Some of the fiercest criticism came from India, where the antiglobalization movement campaigned tirelessly against the encroachment of biotech crops on the country's vast rural economy. It was against this backdrop that peasant farmers from the western state of Gujarat sent the debate over biotech crops in a startling new direction with a hack that no one saw coming.

During the 1990s, Indian journalist P. Sainath began documenting a wave of suicides among desperate peasant farmers crushed by debts in the country's poorest regions—a wave that has yet to subside. The toll was especially high among farmers of cotton, a water-intensive crop that often fails during drought years. As the number of suicides grew into the thousands, and news media outside the country began to pay attention, arguments over what and who were at fault intensified. Sainath blamed India's embrace of the World Trade Organization, which he said led to policies that pushed subsistence farmers away from food crops and toward commodity crops that carried higher costs and risks. Centrists faulted allegedly corrupt government bureaucracies that they said funneled unsustainable subsidies to farmers while basic infrastructure like irrigation crumbled, leaving growers bereft when the money ran out. Others said India was simply leaving its rural poor behind as the country embraced a high-tech future of Western-style free-market prosperity.

At the center of the debate was the cotton itself. As in much of the world, genetically modified crops had stirred up angry opposition that had become part of mainstream political debate in a way that never happened in the United States. Politicians tapped into worries that genetically modified cotton in particular was both dangerous in itself and a symbol of U.S. technological hegemony. Antiglobalization activists painted the approval and spread of genetically modified cotton seeds that had been engineered by Monsanto as a move toward neocolonialism. The activists argued that Indian peasants would become indentured not to the local raja but to a Western multinational that would use patents to control their lives. Instead of using traditional seed varieties bred and saved year after year, opponents said, the licensing fees that would give farmers the right to plant Monsanto's pest-resistant seeds sent already desperate growers deeper into debt than they were already.

Monsanto has called those allegations "sensational" and "speculative" and claims its genetically modified seeds have allowed Indian cotton farmers to prosper.

The seed that fueled the conflict is known as Bt cotton, named after a naturally insect-resistant bacteria, *Bacillus thuringiensis.* The genes from *Bacillus thuringiensis* that code for the bug-repelling protein are spliced directly into the cotton genome of Monsanto's seeds. The modified seeds allow the cotton crop to ward off the pink bollworm, a ubiquitous scourge of Indian cotton farmers. Supporters of Bt cotton argued that wide adoption would let Indian farmers increase yields while drastically cutting down on the use of pesticides. Opponents condensed their arguments into one succinct message: Bt cotton would mean trading traditional peasant agriculture for corporate servitude.

In Gujarat, something else happened that neither side expected. In 2000, a respected Gujarati seed company, Navbharat, began selling a hybrid cotton seed known as Navbharat 151. The farmers who planted those seeds did not attract any special attention until the next year. In 2001, a bollworm outbreak ravaged the Gujarat cotton crop. But not that grown from Navbharat 151. As Washington University anthropologist Glenn Davis Stone put it in a study of what happened next, these unspoiled plants amid the ruined fields "raised eyebrows."

At the time Navbharat 151 went on the market, the Indian government had not yet approved the planting of any form of genetically modified cotton. When tests quickly confirmed the Navbharat 151 contained the same Bt gene that endows insect resistance in Monsanto's seeds, the government cracked down. Company officials were charged with violating environmental regulations, since the seeds had never been approved. The government ordered the crops destroyed.

But Navbharat never faced any legal blowback for selling seeds containing the patented Bt gene without getting a license from Monsanto, because at the time genes were not patentable under Indian law. Navbharat claimed not to know that the Bt gene had been bred into the seeds. D. B. Desai, Navbharat's owner, said his seeds must have been contaminated by test plots of Monsanto's transgenic plants. At the same time, he sought a license from Monsanto when his seeds' secret weapon was revealed. Monsanto did not grant the license, but the Indian government did approve three varieties of Monsanto Bt cotton seeds in time for the 2002 growing season.

By then it was too late for Monsanto, at least in Gujarat. Despite the government's public display of outrage, the Navbharat crops were not destroyed. And although Desai and his company could no longer sell their allegedly pirated seeds, the seeds themselves survived. So Gujarati farmers did as they had always done with successful crops: They saved the seeds to plant again.

In the United States, Monsanto investigators travel the country's rural byways to monitor for any unlicensed use of the company's patented transgenic crops. Growers who are discovered cultivating Monsanto plants without paying the required licensing fee can face harsh civil penalties.

Not so in India. As the second generation of Navbharat 151 seeds began to spread, Monsanto had neither the authority nor the infrastructure in the country to keep tabs on every peasant farmer with a few acres in a state of fifty million people. And Indian politicians were hardly eager to act as the long arm of Monsanto, keeping farmers from getting the high yields that have helped Gujarat cement its place as the biggest cotton-producing state in India. Instead, the contraband genetically modified seeds began to spread. Farmers began passing around what Cornell political economist Ronald Herring has called "stealth seeds."

Though biohackers doubt the value of building walls around intellectual property and question the motives of corporate biotech, they still might not have turned the farmers of Gujarat into folk heroes if not for their next gambit. After all, using Monsanto's seed technology without the company's permission is about as innovative as downloading pirated songs off the Internet. Some might see revolution in a sort of adolescent defiance of authority, but pure piracy benefits no one except the individual end user.

Yet what happens when the individual downloader takes the stolen tracks and remixes them to create a new sound? And what happens if that new sound grips a subculture of listeners like nothing they had ever heard before? The farmers of Gujarat did more than hijack Bt cotton, plant the seeds in the ground, harvest the crop, and repeat. Instead, they did what they have always done with regular seeds. They took one variety of seed that they believed did one thing well and crossed it with another variety that did something else better. What they ended up with was a better seed than anything they had come up with before on their own—and also better for them than anything Monsanto had built besides.

According to Herring and others, farmers in Gujarat have created a "cottage industry" around crossbreeding seeds descended from the original Navbharat 151 crop. Like the horticultural savants who have bred an endless number of popular, potent marijuana strains, Gujarati farmers mixed and matched cotton varieties in search of the hardiest, highest-yielding crop. Trust in pure second-generation 151 seeds was low, as it has always been with any seed unchanged since the previous season. As all farmers know from experience, nature over time tends to find the weaknesses in any crop strain. Pests and blight develop a taste for a particular variety. Climate pushes a crop to the breaking point. The time-honored way to counteract these inexorable ecological forces is to cross two strains to create a new variety that shakes up the natural order enough to buy the crop some time for another season.

The stealth seeds crossed with more conventional varieties included not only the Bt gene but also natural mutations from other nontransgenic strains that offered some advantages particular to the climate and ecology of Gujarat. Herring reported that without any official sanction behind them, Gujarati farmers and merchants dealing in stealth seeds had to rely on the same dicey webs of reputation and trust that run through any underground economy. Farmers provided the stealth seeds to retailers, some of whom in turn guaranteed to other buyers that the seeds contained Bt. Seed pirates relied on not so subtle marketing ploys like naming their product BesT Cotton Seed, which continued to sell despite the approval of Monsanto's own seeds for use in India.

Gujarati farmers created an assortment of niche crops that meet their particular economic and ecological needs better than anything else out there. What's more, they did this on their own, without anyone else's permission. Whether legal or not, they took control of the technology in their hands and, in a rough-edged way, reverse engineered it, hacked it, and forcibly open sourced it. They dragged Monsanto's seeds into the public commons, and in so doing cleared the path that led to the innovations that mattered to them. They combined the latest biotechnology with the most ancient practices for domesticating life to create something new. According to Herring, "Stealth transgenics are [being] saved, cross-bred, repackaged, sold, exchanged, and planted in an anarchic agrarian capitalism that defies surveillance and control of firms and states."

And why would the farmers rebel in this way, creating underground seed markets that could unleash, depending on who you ask, environmental catastrophe, Monsanto attorneys, or the wrath of the Indian criminal justice system?

"Farmers pursue stealth seeds because the technology is affordable and divisible; the genetic roulette they enable has not been a powerful deterrent," Herring wrote. ". . . Farmer-bred 'Robin-Hood' Bt seeds may be spreading faster than officially sanctioned seeds from 'monopolist' Monsanto, as they are cheaper, and often give better results." How much cheaper? Herring says farmer-bred hybrid transgenic varieties can cost less than a third of the price of Monsanto's licensed Bt seeds. As a result, illegal seeds are outselling legal seeds by as much as ten to one.

In the process of liberating the seeds from both corporate and government control, the farmers also liberated themselves from both sides of the debate over genetically modified crops. On the one hand, the farmers were willing to disregard the warnings of antitransgenic activists—and the Indian government's ban on transgenic cotton—when they found that transgenics worked better for them. On the other hand, the warnings about becoming serfs ruled by Monsanto turned out to be moot, because they never paid the company for the seeds in the first place. They were never co-opted because they opted out. It was a radical gesture that nicely fit the anticorporate agenda, except for the embrace of the very genetically modified seeds that activists decried.

The Gujarati farmers' example inspires biohackers who believe that do-it-yourself biotechnology's true world-altering potential will not be unleashed until it reaches the hands of the have-nots. Their use of biotechnology as a tool of self-determination seems to undermine both sides of the political debate over transgenic farming. Instead of empowerment through defiant embrace of tradition (or government subsidies), the farmers chose Bt cotton. At the same time, they rejected the supposed benefits of taking part in the aboveground global economy, choosing its technology but rejecting its rules.

This is not to say that biohackers expressly condone the piracy of intellectual property. But the innovations the Gujarati farmers made when they began remixing their seeds has convinced DIYers like Guido Núñez-Mujica that the patent system meant to protect innovators' profits has slowed the pace of progress.

"The farmers of Gujarat will be more and more the face of distributed biotechnology, not the geeks in the Bay Area. Their basic needs are already solved," Núñez-Mujica says. "It's the farmers in Gujarat and in Venezuela where this is going to mean the most."

CHAPTER 8

My Life

In March 2010, a federal court ruled that life was not for sale. At least, that is how an unlikely alliance of bicoastal progressives and heartland religious conservatives hailed U.S. District Court judge Robert Sweet's ruling against the U.S. Patent and Trademark Office, a ruling that threatened to turn the biotechnology industry upside down. Since the first gene patents were awarded some three decades ago, the federal patent office has granted patents on about 20 percent of the human genome—thousands of genes in all. For years, the rationale behind such patents came down to the idea that the scientists who had done the work of isolating specific genes should get to reap the benefits of exploiting that information for profit.

In the case brought before Judge Sweet, Salt Lake City–based Myriad Genetics was defending its patents on two genes that can have diagnosable mutations well-known to signal a heightened risk of breast cancer. Myriad took advantage of its patent to license tests that diagnosed these mutations in the two genes—BRCA1 and BRCA2. The test cost more than $3,000 per patient. Positive results compelled many women to make a wrenching decision: Should they undertake preemptive treatment, including having one or both of their breasts removed to avoid the cancer risk at a time when they did not have cancer?

Sweet found in his ruling that DNA was unique among biological chemicals in its primary function, which he said was to encode information. Identifying which information was encoded in specific snippets of human DNA was, in the judge's words, akin to discovering "laws of nature," specifically "those that define the construction of the human body." Newton never owned gravity. The Manhattan Project never paid Einstein a licensing fee for $E=mc^2$. Following the judge's logic, just finding out what breast cancer genes do does not mean that only you have a right to act on that knowledge. They are still nature's handiwork. Myriad's loss of its patent forced countless biotech CEOs into somber huddles with their attorneys. They fretted that the court's decision would allow competitors to profit off their hard-earned discoveries. Meanwhile, civil liberties groups and open-science advocates celebrated what they saw as an end to the corporate lockdown on knowledge that they believe should belong to all.

The controversy over gene patents has existed as long as the biotech industry itself. In 1972, University of California, San Francisco (UCSF) microbiologist Herbert Boyer and Stanford researcher Stanley Cohen met at a Honolulu deli during a conference. Over pastrami sandwiches they conceived an experiment that would launch the modern biotech industry. Back in California, the pair successfully inserted frog DNA into bacterial cells that began reproducing themselves, including the inserted frog gene. The end result became known as recombinant DNA, though journalists have always preferred to talk of spliced genes.

Boyer and Cohen published three papers on recombinant DNA in 1973 and 1974 that rocked the scientific community. Before long, the idea of genetic engineering had filtered into the media, often with menacing overtones. Scientists and the public responded with both the fear and hope that have greeted so many major scientific discoveries. Some wondered whether this newfound ability to manipulate genes could lead to dangerous perversions of nature or microbial Frankenstein's monsters. Under the headline DOOMSDAY: TINKERING WITH LIFE, a 1977 article in *Time* quotes Caltech biology chairman

Robert Sinsheimer: "Biologists have become, without wanting it, the custodians of great and terrible power. It is idle to pretend otherwise." Others foresaw great promise: Could this same ability finally unlock the cures to the most intractable diseases? If the latter, the profit potential for recombinant DNA could be huge. That potential was lost on neither Boyer nor Stanford University.

Boyer harnessed his discovery to cofound Genentech with the backing of venture capitalist Robert Swanson in 1976. At the time, few biologists in academia ventured into the corporate realm. Genentech's massive success helped change that dynamic. In its early years the company produced the first human protein made in a microbe, cloned human insulin, and created synthetic human growth hormone. In 1980, Genentech went public and saw its share price surge from $35 to $88 in less than an hour on the market. The sale raised $35 million and made Boyer's fortune.

The same year as Genentech's initial public offering (IPO), the U.S. Patent Office granted Stanford and UCSF the first-ever patent on the recombinant DNA technique. The Cohen-Boyer patent became one of the twentieth century's most famous pieces of intellectual property—and one of the most lucrative. Over the next twenty-five years, the discovery would bring the two universities a combined $255 million in licensing fees, as recombinant DNA became the foundation of the modern biotechnology industry. Meanwhile, the patent itself became the gold standard for academic research institutions seeking to reap revenue from their scientists' discoveries.

The way to the Cohen-Boyer patent was paved by the June 1980 U.S. Supreme Court ruling *Diamond v Chakrabarty*. The 5–4 decision found that a bacterium genetically engineered to devour crude oil was a product of human ingenuity and therefore patentable, regardless of its status as a living thing: "Anything under the sun that is made by man," in the ruling's most famous phrase.

Two weeks after the Cohen-Boyer patent was granted, the Bayh-Dole Act went into effect. Named after its bipartisan U.S. Senate cosponsors, the law allowed universities, small businesses, and nonprofits to retain the rights to inventions developed with federal government backing. Until then, rights to these inventions, devised with government funding, had reverted to the government. Universities have prospered financially as a result of the law. But critics have long contended that the profit motive conflicts with universities' core mission of advancing knowledge for its own sake and that the law invites corporate interference in research priorities.

Over time, the biotech business model has come to depend at its core on the protection of intellectual property. Discoveries yield patents. Patents yield products. Products yield profits. Like software, nearly all of the costs in the development of biotech products come up front during research and development. To recoup that investment, companies must stay vigilant to keep competitors from filching that hard-won knowledge and infusing their own products with its value. The pills and seeds these companies make are not worth anything in themselves; they are not commodities like gold or oil. They are rather vessels that transport the microscopic physical embodiment of the knowledge and work that led to their creation. And they make some people very rich.

Biopunks have problems with this model. Not so much the making money part—financial gain in itself is of little concern to them one way or another. But the same do-it-yourself biologists who claim that expensive equipment, chemicals, and facilities have allowed large institutions to monopolize biotechnology also see that monopoly extending to the ownership of knowledge. In their eyes, the traditional model of intellectual property keeps knowledge barricaded behind the castle wall. Great discoveries are protected by powerful gatekeepers such as patent attorneys and university licensing offices while the serfs—the rest of us—have no real access.

Biopunks want to tear down the walls and throw open the gates. Not to pillage or destroy, as they see it, but to carry knowledge aloft on their shoulders into the village and the public square. They see themselves as the ones who will nurture and tend and, perhaps most important, play. They see scientific knowledge not as a source of power that they can exploit for personal gain. Rather, they would like science to act as a liberator. In their vision, knowledge walled up will only ever benefit the few, while knowledge set free will empower everyone.

According to conventional wisdom, three things will attract investors to your biotech company: intellectual property, successful clinical trials, and hype. As a reporter who has covered the industry, I have been on the receiving end of the hype machine. You might think that an industry charged with solving the greatest health problems of our time would brim with stories of dynamic companies and scientists taking big risks and making bigger discoveries. Maybe, but panning for those nuggets out of an avalanche of press releases can be discouraging. You sometimes get the feeling that many companies exist not to push new science but to move markets by spinning incremental advances as major breakthroughs.

This works, because some investors will always be willing to take a chance on slim data in hopes of replicating the success of Genentech. By the time Swiss pharmaceutical giant Roche bought the company outright in 2009, Genentech was valued on the New York Stock Exchange at more than $85 billion on annual revenues of more than $13 billion. At the time of the sale to Roche, Genentech had ten drugs in its portfolio, including several of the world's top cancer-fighting drugs. Its top seller, Avastin, brought the South San Francisco–based company nearly $2.7 billion in domestic sales in 2008.

In the world of corporate biotech, success is measured by Avastin. First approved for use in 2004, Avastin battles cancer by inhibiting the growth of blood vessels in tumors. The FDA has approved it for use against colorectal, breast, brain, kidney, and lung cancers. The drug can cost up to $100,000 per year per patient. The start-up that discovers the next Avastin would make its backers wildly rich. Yet even Avastin only prolongs life in patients by a few months on average. The company that does Avastin one better—increasing survival by six months or a year, for example—would make everyone involved into instant titans. For the company that finally cures the disease, the rewards could mint the next Bill Gates.

In short, the potential rewards in a venture capitalist's risk calculation are huge. Traditional corporate biotech also appeals to investors in its resemblance to the software industry. In developing a new biotech drug, much like building the next killer app nearly all the main costs come from research and development. While mass-producing Avastin is more complex than distributing new copies of Windows 7, the cost of the drug does not come from the expense of brewing vats of genetically engineered cancer-fighting chemicals.

Instead, to understand where the best money in biotech is spent, look no farther than Genentech's sleek main campus. Some three dozen buildings cluster in tranquil isolation along the western edge of San Francisco Bay. Workers can take in the views as they stroll along DNA Way, the campus's main thoroughfare, or just hop a shuttle to make their next meeting. Employees enjoy top-notch child care and dining, concierge service and free coaches to and from company headquarters for commuters across the San Francisco Bay area. Researchers are allotted 20 percent of their work time—one day a week—to pursue independent projects of their own choosing (one of these led to the development of Avastin). Like Google, Genentech strives to create a cocoon of comfort to nurture its most important asset: its human idea-generators. Pharmaceuticals do not create value for Genentech. The people do.

And if just one of those people has the one insight, the one hack, the one "Eureka!" out of which a new therapy emerges, the return on investment quickly mounts. Squeezing innovation from the ether takes cash, the conventional investment wisdom goes. Once that hard thinking is done, investors hope the drugs themselves will mint money.

Yet as with Google, for every Genentech in the biotech industry, hundreds of start-ups rise and fall with little fanfare. If free sushi, shuttle buses, and buckets of money were all that were needed to concoct blockbuster drugs, cancer by now would have gone the way of polio.

Instead, what the biotech industry has managed to eradicate more than anything else so far is cash from investors' pockets.

More than forty years after Genentech was founded, biotech industry analysts reported in 2009 that the industry as a whole had finally turned a profit the previous year. For four decades prior, investors poured tens of billions of dollars into countless companies in pursuit of the next big biotech jackpot. Serious medical advances followed, but the familiar list of diseases that still lack seriously effective treatments shows in how many ways biotech has failed to live up to its early promise. Alzheimer's, diabetes, and multiple sclerosis. Parkinson's, Crohn's and the common cold. More cancer treatments exist than ever before, yet patients at the most advanced treatment centers in the world still endure brutal surgeries, radiation, and the torture of chemotherapy.

As the body of basic science describing the nature and origins of disease expands, calls have grown louder in recent years for a better explanation of why so much knowledge and so much money have led to so few cures. A common response is that biology as a scientific discipline is hard, a plain fact that cannot be overstated. But biopunks contend that the system within which scientists practice that discipline must bear some of the blame. Much scrutiny has fallen on the biotech industry's all or nothing, lotterylike business model, which favors the pursuit of the next blockbuster drug above all else. Historically, the search for the next big score comes back to the protection of intellectual property, which critics say has become a core function of the biotech industry. Because patents are prized, companies hide their discoveries until their patents are approved. Biopunks argue that in such a closed system, the wisdom of the crowd is thwarted. The best minds are discouraged from taking a crack at the best data, they say. An intellectual property system designed to spur innovation by allowing inventors to profit off their inventions has become in biopunks' eyes a high-stakes game of low-stakes progress. As lawyers and investors bicker over rights, real advances stall. Biopunks want a different business model to resurrect what they see as a more authentic spirit of invention, the kind that animated Ben Franklin and Thomas Edison, the mythical heroes of American innovation. Biotech needs someone to fly a kite in a lightning storm, but biopunks believe the obsession with intellectual property is killing the breeze.

Andrew Hessel is a particular kind of character you encounter in the tech world, especially in the Bay Area. Charismatic and articulate, he favors black button-down shirts and jeans. His red-framed glasses set off his salt-and-pepper bedhead. Even when appearing before a large group, he often needs to shave. He talks about the time he dropped out to Thailand for a year to think about things. He jets around the country twenty-six weeks out of the year giving talks, having meetings, setting projects in motion. Based on appearance alone, he could be a record producer or work in creative at a boutique ad agency. Instead, he is the closest thing I have seen to a biotech hipster—a professional life sciences provocateur.

Hessel believes that bioengineering can solve the world's problems and tries to inspire others to feel the same. He sees the exponential advance in the power of information technology as a mere prelude to the same kind of expansion of power to manipulate DNA. But Hessel's techno-optimism stops at drug development. He thinks drug development sucks.

During the sixty years or so that computers went from a roomful of vacuum tubes to iPhones, Hessel laments that the pace of drug development has never quickened.

"The process we've been using is so complex, with so many stakeholders, that we've never been able to accelerate," Hessel says.

Hessel worked at one of the nation's largest biotech pharmaceutical companies for seven years during the 1990s. "At the end I had to walk away, because it wasn't innovating. We made two drugs, which made us billions of dollars. In seven years, we never made another drug," despite spending $1 billion on research. "That's a really bad return on investment. And it drove me crazy."

But Hessel would not be a biopunk if he did not have a hack at hand. Except in this case, his solution is not silicon-based. It's social.

Hessel believes that drug development will only accelerate if drugmakers streamline the number of stakeholders to one: you. To that end he founded the Pink Army Cooperative, which he calls the world's first drug development co-op. A share of the company costs $20. Once you buy it, he says, you own a pharmaceutical company.

His argument goes like this: Cancer ultimately is not complicated. Whatever form it may take, cancer itself is simply a corruption of our genetic material. If DNA provides our bodies with operating instructions, cancer is a misspelling, which if left unchecked causes bad code to be spread throughout the network—ourselves. With the latest tools we can see those misspellings with a high degree of resolution. Even then, however, treating cancer remains difficult, because the tools we have to target the disease lack the same hi-res precision. Current drugs do not differentiate well enough between the correct and incorrect genetic spellings. Instead, Hessel says, cancer treatment takes a "nuke it from orbit" approach. In the end, he says, you may slow or stop the cancer. However, "you feel like crap because you're getting an optimized poison that's the best we've got from the last century."

Hessel does not see the science itself as the main problem in getting past the old ways of treating cancer. He faults the system in which the science is done. "The hardest part about cancer these days is that change is hard. There's a lot of people that have been trained to think about cancer in a certain way," he says. And that way is to make drugs that work for everyone. "The clinical drug development pipeline is tuned to make things like blockbuster movies."

This is irrational when dealing with a personalized disease such as cancer, he says. "I started to think, what if I got cancer tomorrow? It's not hard. I smoke. I would want something absolutely specific to my cancer. I don't want to know how it's going to affect you." He would rather think about cancer drug development in terms of *Wired* magazine editor-in-chief Chris Anderson's idea of the "long tail." In the same way that companies like Amazon and Netflix have prospered by figuring out how to deliver a multitude of products to a multitude of niche markets, a company should be able to economically deliver as many cancer treatments as there are cancers. Instead of attempting a one-size-fits-all approach to cancer drugs that can make one company a lot of money but not serve any one patient all that well, Hessel wants to create a system where every patient is their own control: Either the treatment works or it doesn't.

This is the driving idea behind Pink Army. As a co-op, Pink Army has no incentive to seek the big score for shareholders. Instead, everyone owns an equal piece, and each person can get involved as much or as little as he or she wants. If a treatment works for one member, then that information gets fed back into the system. Another member tries it. It works or it doesn't. From each of these experiments, a greater understanding is gained. Through this open-source feedback loop, combined with the increasing accessibility of tools to research the genetics of cancer on a refined level, Hessel hopes to break what he sees as the boom-bust cycle of large-scale drug development. Of course, such self-experimentation is always risky, but, as Hessel points out: You have cancer. He also doesn't worry much about regulators. When everyone involved owns the operation equally, and everyone is conducting a clinical trial of one, who do you fine? What exactly do you shut down?

So far, Pink Army is more a concept than an actual co-op. Hessel needs more buy-in to achieve the critical mass of money and minds needed to do meaningful research. He believes if one charismatic patient steps forward to become her own experimental test subject, enthusiasm could reach the tipping point needed to transform Pink Army from a provocative idea to a practical new approach to drug development.

"We are trying to be the Linux of cancer. We need people to invest in themselves," Hessel says. "I charge twenty dollars per share. It's the price of the pizza. If you are not going to invest the price of a pizza, you might want to rethink how strongly you care about cancer."

CHAPTER 9

Ladies and Gentlemen

Cracking open the book of life has always carried a whiff of danger. Since the earliest human hunters began tracing the habits and habitats of their quarry, the study of living things has brought with it the possibility that the object of your inquiry could kill you. The prudent fear stirred by this very real threat, whether a wooly mammoth or an Ebola virus, has led to the tightly controlled safety protocols governing the practice of biology today. Those protocols are so pervasive and sensible that most biologists cannot imagine and would not see the point in conducting their work any other way. Yet biology as a field would not exist if its early innovators had not put themselves at great risk to make the discoveries that would push the field forward.

In the portrait, her skin looks smooth as glass. She wears a jeweled Ottoman headdress and an embroidered great coat. A knife juts from a thick woven belt, though its bare blade looks less sharp than the glint in her eye. With a mere cock of the eyebrow, Lady Mary Wortley Montagu conveys all the wit, intellect, and arrogance of the British empire at the dawn of its ascendancy. Nowhere does the unnamed portraitist hint at the scars that blighted the face of the woman whose beauty inspired the eighteenth century's most famous English poet, Alexander Pope, to write: "In beauty, or wit, / No mortal as yet / To question your empire has dar'd."

Around the same time that her husband became British ambassador to the Ottoman empire, Lady Montagu contracted smallpox, a disease as common to her era as cancer or heart disease to the early twenty-first century. About 30 percent of those afflicted by the disease died. Those who survived the fever, vomiting, and pustulant rash, like Lady Montagu, spent the rest of their lives disfigured. Yet even in a society as captivated by appearances as her own, she did not retreat from public life. Instead, she became the driving force behind an effort that led to the greatest public health revolution in history.

For nearly the entire existence of the human species, the practice of medicine and the pursuit of medical knowledge have been ugly. Physicians had little to work with beyond the direct observation of their five senses. Even then, such basic knowledge as the purposes of the heart, stomach, brain, and lungs defied understanding for millennia. The trail of exploration and experimentation that led to the detailed understanding of physiology and health we take for granted today is littered with suffering and risk.

As biopunks challenge the prevailing notions of who gets to participate in this age-old quest to lift the burden of sickness and disease, the stories of an earlier pair of scientific risk takers remind us of just how recently contemporary notions of medical professionalism came into existence. Vital discoveries, such as the invention of vaccines, occurred without any of the insights afforded by modern molecular biology or any of the safeguards provided by modern standards of institutional research. Biohackers like Andrew Hessel claim that those same institutions have become twisted by irrelevant incentives and unnecessary restrictions in ways that steer scientists away from the mission of bold scientific innovation. Looking to the great discoveries of the past and the people who made them, it's tempting to conclude that professionalization has been the enemy of innovation.

But what would it mean to return to an era of aristocratic amateurs like Montagu and gentleman country doctors like Edward Jenner, the father of modern vaccination? Fueled by raw curiosity and a desire to end suffering, these undisputed scientific pioneers changed not just the face of medicine but the course of human history. And they did it themselves because they had no other choice—there was no scientific establishment to rebel against in any meaningful way. Yet the stories of what they achieved do not offer easy answers about how to weigh the possible benefits of important discoveries versus the potential of the work leading to those discoveries to do harm. Their enthusiasm was not hampered by anything like the strict legal and ethical codes that govern much scientific practice today. But where that enthusiasm took them also forces some unpleasant questions about whether the good of the many can ever outweigh the price of harm to a few.

Not long after her own bout with smallpox, the disease killed Lady Montagu's brother. She arrived in Turkey a typically tragic victim of the so-called speckled monster. Yet within a few weeks, she had witnessed a practice that left her amazed and transformed. Her detailed description of the seemingly primitive medical procedure in a letter to a friend has become a classic passage in the history of medicine:

> A propos of distempers, I am going to tell you a thing, that will make you wish yourself here. The small-pox, so fatal, and so general amongst us, is here entirely harmless, by the invention of engrafting, which is the term they give it. There is a set of old women, who make it their business to perform the operation, every autumn, in the month of September, when the great heat is abated. People send to one another to know if any of their family has a mind to have the small-pox; they make parties for this purpose, and when they are met (commonly fifteen or sixteen together) the old woman comes with a nut-shell full of the matter of the best sort of small-pox, and asks what vein you please to have opened. She immediately rips open that you offer to her, with a large needle (which gives you no more pain than a common scratch) and puts into the vein as much matter as can lie upon the head of her needle, and after that, binds up the little wound with a hollow bit of shell,

> and in this manner opens four or five veins. . . . Every year, thousands undergo this operation, and the French Ambassador says pleasantly, that they take the small-pox here by way of diversion, as they take the waters in other countries. There is no example of any one that has died in it, and you may believe I am well satisfied of the safety of this experiment, since I intend to try it on my dear little son. I am patriot enough to take the pains to bring this useful invention into fashion in England, and I should not fail to write to some of our doctors very particularly about it, if I knew any one of them that I thought had virtue enough to destroy such a considerable branch of their revenue, for the good of mankind.

Lady Montagu's description was inaccurate on one important count. According to estimates from the National Institutes of Health, between 1 percent and 2 percent of patients deliberately infected with a small amount of smallpox at the time still died from the disease. Still, what she had witnessed was an ancient form of inoculation that increased chances of survival more than tenfold compared to those of a full-blown case of the disease. When Montagu gave birth to her daughter in Turkey following her bout with the disease, embassy surgeon Charles Maitland summoned Dr. Emmanuel Timoni, a prominent Turkish physician, to the delivery. Timoni had tried a few years earlier to distribute a book in English on "variolation" (the name given the procedure after the virus that causes smallpox) but found little interest among British physicians. Timoni saw Montagu's scars and knew she would want to protect her child from the scourge that had so ravaged her. He asked if he could variolate her only son.

So great was her confidence in what she had seen and heard about the procedure while in Turkey that she followed through. The boy, five-year-old Edward, would grow up to become an author, member of Parliament, prolific traveler, and infamous rogue who toward the end of his life began dressing much as his mother had in the portrait from her early days at the Turkish court. During his more than sixty years, scandal often trailed Edward as he crisscrossed Europe and the Middle East, but the disease that had scarred his mother and killed millions never caught up with him. In 1776, he died in Padua from choking on a fish bone.

Convinced that she had successfully shielded her son from one of humanity's great scourges, Lady Montagu returned to England and promptly had Maitland inoculate her four-year-old daughter in front of the court of George I. The demonstration promptly gained him permission to experiment more widely. His first test subjects were six prisoners promised favorable treatment if they participated. None died, and later exposure to smallpox among some proved they had become immune. Maitland followed up with experiments on orphans. His apparent successes, or at least lack of blatant failures, convinced the British upper classes of the validity of variolation. Europe's elite soon followed.

Edward Jenner was not born to that elite. The son of a clergyman, Jenner was born about 120 miles west of London in the country town of Berkeley in 1749. An orphan by age five, his older siblings sent him to free boarding school when he was eight. A smallpox outbreak at the school led the staff to variolate the children, including the young Jenner.

At the time, a fully formed germ theory of disease was still more than a century away. Variolation worked, but British physicians had no idea why. Still grounded in medieval theories of physiology and illness, English doctors forced anyone awaiting variolation to endure six weeks of bleeding, fasting, and purging beforehand.

The process left patients sickly, thin, and weak—in other words, less suited to fight off the smallpox infection they were about to receive. Nearly fifteen years after Lady Montagu had introduced variolation to the nation in 1721, only 850 patients had undergone the procedure, mostly because of the suffering promised by what came before. When he was inoculated, Jenner and the other children at his school had to endure the brutal six-week preparation themselves.

The procedure left the boy suffering from what sounds today like posttraumatic stress disorder—insomnia, severe anxiety, even auditory hallucinations.

With that early horror burned into his memory, Jenner switched schools and began tinkering with the natural world. He raised dormice and collected fossils, which interested him more than his mandatory studies of Latin and Greek. In a story still painfully familiar to geeks of all kinds, his idiosyncratic interests combined with his lack of conventional academic success meant Jenner could not likely follow his older brothers to an elite university, in this case Oxford. Instead, he started down a path that seemed likely to lead him to a major loss in social status compared to university-trained scholars of theology and the classics. He became a surgeon.

Meyer Friedman and Gerald Friedland write in their account of Jenner's life: "Although changes were beginning to occur in the British medical system, the division between physicians and surgeons still existed. Surgeons, who were much less educated, acquired their medical knowledge through apprenticeship rather than academic work at one of the universities." Jenner began his own apprenticeship at thirteen, training with a country surgeon for six years, during which he first heard the tales of milkmaids who declared themselves immune from smallpox because they had earlier contracted cowpox, a far milder disease that causes only superficial illness in humans.

Jenner's ranging mind led him to London at twenty-one to study at St. George's Hospital under John Hunter, one of the top surgeons of his day and a tireless gentleman hacker who taught Jenner the scientific method. Though in London for only two years, Jenner's talent for biological observation led famed naturalist Joseph Banks to hire him to catalog the specimens the latter had brought back from his voyage with Captain James Cook to the South Pacific. Afterward Jenner returned to Berkeley, where he was born, to work as a country doctor and gentleman scientist.

From a distance, Jenner stayed deeply engaged in the scientific advances of the day via his correspondence with Hunter. He documented the hibernation habits of hedgehogs. He was the first to document through minute observation the hijacking of sparrows' nests by cuckoo offspring, which will hurl baby sparrows to their deaths and allow mother sparrows to raise them as their own. (The sheer viciousness of the practice so appalled some scientists that debate persisted over Jenner's cuckoo findings until a cuckoo nestling was photographed in the act in 1921. Antivaccine activists used the early skepticism over the cuckoo study to smear Jenner long after his death.) He became obsessed with ballooning, launching his own hydrogen balloon twice at the height of late-eighteenth-century Europe's frenzy over humanity's newfound ability to take to the skies.

He also became one of the first scientists to recognize hardening of the arteries as the likely cause of angina, the chronic chest pain that Jenner recognized was the signature symptom of coronary heart disease. Again, Jenner made his discovery through careful observation and quite literally getting his hands dirty. While performing an autopsy, he was cutting through the heart of a patient who had died while suffering chest pain. His knife struck the clogged arteries, which he described as "bony canals."

Each of these examples of Jenner's inquiries calls attention to the kind of mind able to feel its way toward one of the greatest advances in medical history. He let his imagination roam. He did not isolate himself in an alleyway of specialization. He relied on powers of observation and intuition. He watched carefully, tinkered, and cared little for orthodoxy. In all of these, Jenner fits the mold of the heroic outsider inventor that fuels a particularly Anglo-American mythology of technological progress.

The story of Jenner's development of the first widely used vaccine, after he was inspired by the "folk tales" from the milkmaids that turned out to be true, has become its own kind of scientific folk tale, a beloved yarn repeated over and over until its original power dissipates into sentimentality. Stripped of their romance, the bare facts reveal the halting, prosaic, ethically questionable nature of Jenner's accomplishment. He took risks that could have sent him to jail were he working today, yet he is canonized as the one man in history who may have saved more lives than any other.

Jenner was not the first to recognize that weaker poxes, such as cowpox and swine pox, might afford some kind of immunity against smallpox, just as small amounts of the actual pathogen would. More than thirty years before he administered the first smallpox vaccine, Jenner heard a paper presented by a Dr. Frewster (his first name is lost to history) at the London Medical Society raising the possibility of cowpox's protective properties. In 1774, a Dorset gentleman farmer named Benjamin Jesty, desperate to protect his family from smallpox, chose to believe the milkmaids' tale and exposed his wife and sons to lesions from the udder of an infected cow. The disease passed them by. Still, Jenner appears to be the first person who pursued the idea of a vaccine with any scientific rigor or sense of the public health implications of such a discovery.

Like Lady Montagu, Jenner chose his son, Edward, Jr., as his first test subject. While the boy was still an infant, Jenner inoculated him with swine pox taken from the lesion of Edward, Jr.'s nurse, who had contracted the infection. The child developed his own lesions, and a few weeks later Jenner variolated him with smallpox. The boy showed no sign of infection. Before his son had reached his third birthday, Jenner had variolated him three times to see how long the immunity lasted. He never developed more than a mild reaction. Tragically for Jenner, medical science was far from developing effective treatments for tuberculosis, which killed Edward, Jr., at twenty-one.

After his experiments on his son, Jenner set his vaccination experiments aside for several years. In 1796, he decided to try again, this time with cowpox.

Though a scientific understanding of viruses, much less their biochemical structure, would not begin to emerge for another century, Jenner saw that cowpox and smallpox must share something in common that made milkmaids immune. Because cowpox was not harmful to humans, he predicted that infecting the healthy would provide the benefits of variolation without the risk of death from the very disease being eradicated. He chose as his test subject eight-year-old James Phipps, the son of a man variously described in histories as Jenner's gardener or a homeless laborer. Regardless, the relationship was clearly exploitive: Phipps's father was hardly in a position to say no to his employer, and Jenner knew that any mishap would barely make a ripple because of the boy's low social standing.

As with the earlier experiments, Jenner first inoculated the boy with cowpox, which brought on a mild illness. Weeks later, Jenner variolated Phipps with smallpox. No infection developed. Though the Royal Society rejected Jenner's paper on his results in 1797, physicians responded positively to a pamphlet Jenner published privately the next year. Medical historians now see the pamphlet, "An Inquiry into the Causes and Effects of the Variolae Vaccinae, a disease discovered in some of the western counties of England, particularly Gloucestershire and Known by the Name of Cow Pox," as a milestone in the history of science. With its publication, the modern vaccine era began. Today smallpox no longer exists except as varied locked-down pathogens stored in labs and used as a research tool in carefully controlled environments.

Yet to the modern medical consumer accustomed to the dull march of clinical trials and the Food and Drug Administration's guiding hand, the idea of performing an unregulated medical experiment that involves introducing a lethal pathogen into a child's bloodstream based on purely anecdotal evidence appears barbaric. Ravaged as she was by her own physical and emotional battles with the disease, Lady Montagu's decision to variolate her son could hardly be seen as rational or dispassionate, the hallmark traits of true scientific inquiry. Variolation could easily have killed Edward, who may never have contracted smallpox anyway. Meanwhile, Maitland's experiments on prisoners, who are inherently under duress, evoke an appalling practice nevertheless considered acceptable in the United States well into the twentieth century. Jenner also chose children as the guinea pigs for his most important experiments, which would never meet today's standards of informed consent for research subjects.

Still, the examples of Montagu and Jenner serve as an especially raw reminder that medicine has rarely advanced in tidy steps. Truly vital discoveries rarely lack an element of risk, almost by definition. The unknown in the study of medicine could be the thing that cures you or kills you. You play probabilities, but even in the most strictly controlled twenty-first-century settings you only find out if you try. Lady Montagu took the most profound risk a parent can take, a risk that any decent person would consider abuse by today's standards. Yet history views her as a hero. Though not a scientist, her stature and influence led to the eighteenth-century version of a public health policy shift so momentous that its repercussions rippled into the late twentieth century, when the World Health Organization declared smallpox officially eradicated.

It would be easy to credit Jenner's outsider pedigree as the key to why he and not some other more established scientist of his day became the father of vaccination. His broad enthusiasm for different branches of science combined with his lack of an elite academic résumé gives him the kind of cachet prized among twenty-first-century hackers. He did what he wanted. He worked outside the system. Influential critics condemned his most innovative ideas as radical and wrong. And when it came to making the choice of how to push his discovery forward, he did what he had to do. He experimented on children and saved millions of lives.

No self-described biohacker I have interviewed has showed any sign of wanting to return to an era when a researcher had the freedom to make that kind of utilitarian calculation. The desire to experiment on humans at all, at least on any humans not themselves, does not appear to be part of the discussion. The risks they seek to take or even have the technical ability to take have less to do with potentially dangerous science than the potentially dangerous idea of doing science in a way that looks ahead by looking to a deeper past.

The anthropologist and historian of science Christopher Kelty has identified the "gentleman scientist" as one of the iconic figures, along with the outlaw and the hacker, skulking around in the rhetoric of noninstitutional biology. This does not mean that biohackers are openly calling for a return to eighteenth-century safety protocols or ethics. But they share a romanticism about science that drove their counterparts two centuries ago. Scientific knowledge during that era still held the allure of grand mystery. Gentleman scientists wrapped their endeavors in the language of frontiers and secrets and the great lifting up of humanity through discovery. They had epic ambitions and were by necessity generalists; the knowledge needed for true specialization did not yet exist. They worked in laboratories in their homes stacked high with papers and dusty books, cluttered with the beakers and flasks of mad scientist lore, watched by exotic beasts brought back from expeditions to remote, foreboding islands and seas. They immersed themselves in wonder, and in some cases made wondrous discoveries as a result.

Almost by definition, the gentlemen scientists of earlier eras had a more unfiltered relationship to science, medicine, and ultimately their own bodies. In the twenty-first century, as molecular biology has become so much more sequestered by specialization and bureaucracy, the appeal of the gentlemen scientist to DIYers seems obvious.

Yet the reality does not always hold up against the ideal of the retro-futuristic fantasy.

Modern-day biologists will tell you that specialization is crucial to the advance of scientific knowledge. The only way to advance knowledge is to become fluent in all that came before. As science moves forward, what came before continues to stack up ever higher. Generalism may have romantic appeal, but professional scientists no longer see it as practical.

Moreover, the life of a gentleman scientist was not all wonder and discovery. Just as Jenner had few ethical qualms about whom he chose as test subjects, researchers straining toward the birth of modern science and medicine seldom hesitated to experiment on themselves. In his pursuit of an understanding of optics, Sir Isaac Newton stared at the sun in a mirror until he nearly went blind, and shoved a sewing needle as far as he could into his eye socket, noting the spots of color produced when he pushed against the back of his eyeball. British chemist and inadvertent anaesthesia pioneer Humphry Davy deliberately inhaled various chemicals to better understand their properties. Most famously, he would enclose himself in an airtight chamber and huff quarts of newly invented laughing gas (nitrous oxide) and record the euphoric results. Though he often noted the pain-numbing effects of the gas, he never thought to suggest its use in surgery.

John Hunter, Jenner's mentor and confidant, was admired and respected by his scientist peers. Hunter brought scientific rigor to the study of biology and the practice of medicine, earning him recognition as the so-called founder of scientific surgery. Like Jenner, he came from a humble background and became a professional surgeon when such a calling was considered socially inferior. Trained in medicine as a hands-on practice rather than a scholarly subject, Hunter prized experimentation over study. And like other physicians of his day, he did not hesitate to make himself his own research subject.

Hunter wanted to test the hypothesis that gonorrhea and syphilis were two versions of the same disease. To find out, most scholars believe he injected his own genitals with pus from a gonorrhea patient. When he contracted both diseases, he decided that the hypothesis was true. Such was his influence that his false conclusion (the conditions are actually caused by two different germs) set back research into venereal disease by decades. And since no effective treatments for either disease existed until the development of modern antibiotics centuries later, Hunter suffered with both until he died of a heart attack in 1793.

CHAPTER 10

Cancer Kitchen

John Schloendorn is obsessed with death, but on the day I met him in his lab, he didn't look the part. He wore a bright orange polo shirt printed with palm trees, though his pants were black. He is skinny, spry, and quick to laugh, showing off a row of bright white teeth beneath a shock of dirty blond hair. A native of Munich, Schloendorn at once embodies and undercuts several pop culture stereotypes about Germans. His fixation lacks the consuming seriousness of a Heidegger or a Herzog. Schloendorn thinks death is ridiculous—and something really ought to be done about it.

At the time I met him, Schloendorn told me his new company had just received a half-million-dollar investment to pursue his ideas about how to manipulate the human body's own immune system to kill cancer cells. The twenty-eight-year-old scientist sees this approach not merely as another possible advance in treating cancer but as a possible way to cure it. Still, he has greater ambitions.

"Curing death as a goal was always completely self-evident to me. It was just a question of how to do it," he told me, laughing, but also totally serious.

"I had an interest in not dying, and that led to an interest in science as a means to get there," he said. "I don't think that's ambitious. I think life is the basic prerequisite for doing anything, so that's the first that we need to ensure. Then we can think about what to do with our lives."

Schloendorn came to the United States in 2005 after a serendipitous meeting at Cambridge University during a conference for longevity research devotees. The meeting was hosted by Aubrey de Grey, a Cambridge computer scientist turned gerontologist who has attracted both a passionate following and passionate criticism for his ideas about halting the aging process. Schloendorn, fresh from earning his biochemistry masters' in Germany, had the time, skills, and will to start pursuing the hunches he had about how bacteria might be used to fight heart disease. A private spaceflight entrepreneur put up the money, Schloendorn said, and conference attendee Bruce Rittman, director of environmental biotechnology at Arizona State's Biodesign Institute, had lab space.

A Stanford-trained environmental engineer, Rittman has long focused on finding ways to use bacteria to clean up waste. He has developed technology that uses microbes to decontaminate water and is working to use a similar approach to turn biodegradable trash into electricity. Rittman says so-called microbial fuel cells would rely on specialized bacteria engineered into a "biofilm," a living substance within the fuel cell itself. The bacteria would feed on the biowaste, and in the process of metabolizing it, transfer electrons from the waste to the positively charged side of a battery, creating combustion-free energy. Schloendorn said he and Rittman came together over the suspicion that the same approach to cleaning waste from the environment or transforming trash into fuel could be used to rid the human body of harmful substances. Schloendorn wondered if the same techniques Rittman was using to break down toxins contaminating water could flush cholesterol from the bloodstream.

The idea fit well with the antiaging theories of de Grey, long a polarizing figure who has attracted much media attention for his brash claims about the scientific possibility of immortality. He cultivates the look of a mad scientist with his flowing, chest-length red beard and is not shy about offering up provocative, sweeping sound bites, such as "I think it's reasonable to suppose that one could oscillate between being biologically twenty and biologically twenty-five indefinitely." He was the cofounder of the Methuselah Foundation, which encourages research into longevity by awarding its million-dollar M Prize to researchers who discover innovative ways to lengthen life in mice. As a computer scientist rather than a trained biologist, he is a natural role model for biopunks. His lack of an academic credential did not hinder his pursuit of what he saw as a powerful idea. Whether that idea has yielded results remains debatable.

The underlying premise of de Grey's approach to overcoming aging, which he calls Strategies for Engineered Negligible Senescence, or SENS, holds that science has already identified the seven kinds of damage inflicted on the cells and molecules of the human body. This damage does not cause aging but, in de Grey's analysis, *is* aging itself. All the ailments of aging result from this damage. To halt aging and prolong life indefinitely, de Grey argues, scientists should focus on ways to halt and reverse these seven kinds of damage. Only then can humanity reach what de Grey calls "longevity escape velocity." He has many detractors among mainstream scientists who are as outspoken in their derision as believers are in the ultimate logic of his proposals. Either way, he has managed to attract a stream of very smart people—and funding—to keep his dream alive.

The SENS model identifies one form of damage as "intracellular aggregates" or, as Schloendorn puts it, the "junk" that accumulates inside cells. Among these the most familiar and one of the most deadly is cholesterol, the fatty deposits that build up in arteries and cause heart disease, heart attacks, and strokes. With money and Rittman's support, he moved to Arizona to figure out a way to rid the body of this particular junk.

The first outpost of SENS research in the United States was a small corner of Rittman's lab. Because Schloendorn had his own funding, he did not have to worry about pitching in on the lab's main research, which involves environmental contaminants. He was free to throw whatever microbes he wanted at cholesterol to see if any would devour it into harmless by-products. As his work progressed, it attracted the attention of others who believed in de Grey's ideas, some of whom had deep pockets. Money began coming in from around the world, at which point Schloendorn says he began asking for donors to send soil along with cash. He received soil samples from all over the world, he said. Each contained unique species of bacteria he could test to see how they reacted with cholesterol. The project became the basis for Schloendorn's PhD thesis under Rittman even as the money allowed him to move out of the corner of the lab and into a warehouse space "in the middle of the desert," where he set up his own lab.

Among one of Schloendorn's first volunteer lab assistants was Eri Gentry, a recent Yale graduate from an Arizona prison town an hour down the freeway who was looking for focus for her own big ambitions. The meeting turned out to be an important one, both for the pair and for anyone who ever wondered whether it was possible to start a drug company in your garage.

Gentry grew up in Florence, Arizona, a small town halfway between Tucson and Phoenix that is home to a total of nine prisons. Her parents still own the same grocery store where Gentry told me she started working full-time at age five.

Gentry's high school was so poor, she told me, that the building was condemned. They held classes in makeshift sheds in July. In Arizona. While a neighboring school district spent a million dollars to buy new computers, her school had none to replace. The library was the town library. Still, she was a good student. Helping her parents at the grocery store taught her how to work hard, she says. She devoured anything that had to do with science, she says. Any time there was a science fair, she won.

I asked her whether she had been a geek growing up.

"I was definitely different, but there wasn't any room for geeks," she said. "There just wasn't even enough freedom to stand out. So I mostly kept to myself."

Despite her accomplishments, she says, applying to Yale was just a lark. People from Florence didn't go to Yale. They didn't much go anywhere. But she got in. And then she got out of Florence.

In what would become a recurring theme, Gentry set out to over-achieve again after arriving at Yale. She tried to double major in economics and biology but found even she had a limit. She settled on economics, but her interest in science never waned. Nor her interest in doing something world changing.

Gentry graduated in 2006. And then she was back in Florence. Health problems she was reluctant to discuss sent her back to be close to her parents. Yet what appeared to be getting sucked back into Florence's orbit turned out to be another step forward. While she was figuring out what to do next, she volunteered as an assistant in Schloendorn's small corner of the lab in Scottsdale.

It was there that she got her first close look at what she described as the everyday frustrations in the lives of scientists forced to deal with the constraints of academia pushing against what they really wanted to do.

"It's kind of nose to the grindstone, got to get published, got to get approval from the PI [principal investigator] or above," she said. "These are the same people who would light up when they were at a conference, when they were with friends, when they were chatting about 'how could we innovate in this field?'"

In an institutional environment, she says, she saw that enthusiasm getting shut down, not by any one person but by the whole structure of the system. Instead of working together to make or discover something new, scientists were anxious about someone stealing their ideas and making off with their intellectual property. The result Gentry said she observed was a kind of creativity malaise. "If you have no chance of having ownership over something that you have created entirely, why would you do it?"

Instead of allowing the stifled creativity she saw make her cynical, however, Gentry saw an opportunity. If scientists could opt out of a system within an institution in which every scientific decision could affect their career paths, maybe they could spend more time engaging with ideas and less fretting over office politics.

"I wanted to become an advocate for the people who had great ideas but had no way of getting ownership over them," Gentry said. "And I saw [that] a simple way of doing that was providing them with the tools."

In the end, Schloendorn said he and a growing staff of volunteers discovered several microbes that ate their way through the cholesterol. They had similar success looking for ways to stave off macular degeneration, a common disease of the retina that causes vision loss mainly in older people. The lab generated a few patents. Schloendorn says other SENS researchers are now working to turn his discoveries into a cholesterol-fighting drug.

By the time he was wrapping up that research, he had earned his PhD and was ready to move on. He and Gentry had become allies, and they had become restless. They believed that ambitious research could be pursued on its own terms, without the approval or constraints of institutions, and that's what they decided to do. The final catalyst was the plight of a friend and fellow researcher from the lab at Arizona State who was diagnosed with cancer at age forty. Gentry still simmers over an experimental treatment in Germany that she says the friend could not try because the facilities were closed for the holidays—another example in her eyes of an individual's needs undermined by an institutional culture in which something trivial like vacation took priority over life and death. Within months, he was dead. The pair moved to Silicon Valley to a nondescript apartment in Mountain View a few miles down the road from Google's main campus. They had one goal in mind: to do whatever they could to empower themselves so that others did not have to suffer like their friend. And they weren't going to wait for grants or investors or government regulators or the say-so of anyone else to start trying.

Acquiring cancer cells for research takes more than a credit card number and a mailing address. Although cancer cells kept alive for lab research have never proven infectious in humans, biohackers have found that buying them from biological supply companies and getting them delivered requires certification that you belong to an established research institution. The general public does not have easy access to these materials. Do-it-yourself biologists decry these rules, arguing that responsible citizens should have the right to independent inquiry. "Biopunks deplore restrictions on independent research," Meredith Patterson writes in "A Biopunk Manifesto," "for the right to arrive independently at an understanding of the world around oneself is a fundamental human right."

Because resourcefulness is the essence of any do-it-yourself endeavor, Schloendorn was unwilling to let a little red tape get between him and his work. He was at a Subway sandwich shop when he met with a researcher who was willing to share. The researcher left. Schloendorn, alone with his sandwich, could not contain himself. He called Gentry on the phone from his table:

"Eri, Eri, I'm so excited! I have colon and prostate cancer!" he told her, as the other customers stared busily into their drinks.

In Schloendorn's mind, the goal was simple. To raise money, he needed to show he could create the proper conditions for a white blood cell to kill a cancer cell. After getting their attention, he could sell investors on his hypothesis of how to nudge these cells to act when cancer strikes a person. Killing cancer in a petri dish is not the same as fighting it in a living, breathing patient. And Schloendorn was not the first to have the idea that the cells had this power or to try to show it could be done; research into how the immune system might be manipulated to battle cancer has become a hot topic in recent years. But he may have been the first to try to do it in the comfort of his own home. "To blow up the first cancer cell—that's the risk. And so we just went with [the] minimal equipment needed to blow up a cancer cell. And we could do that at the kitchen table."

The setup embodied the DIY tenet that good enough is good enough. They needed a clean bench, a basic piece of lab furniture designed to keep the air around experiments as sterile as possible. Gentry used a plastic storage bin, some plastic sheeting, and a second-hand HEPA filter to build what could have cost thousands of dollars new. Human cells need a carbon dioxide–rich environment to breed. Established labs will typically have tanks of the gas on hand that feed into temperature-controlled incubators. Gentry and Schloendorn got a block of dry ice (solidified carbon dioxide), captured the fumes in Ziploc bags, and jammed their petri dishes inside. To capture the cells in action, they mounted a USB camera on a microscope.

The experiment itself was simple in concept: Expose the cancers to the white blood cells in his and Gentry's blood. The pair watched a video on YouTube of how to draw blood, then stuck the needles in themselves. Schloendorn purified the blood to get just the white blood cells he needed. He cultured colonies of cancer cells—colon, prostate, and HeLa (pronounced HEE-la) cells. These last are perhaps the most famous and ubiquitous cell line in all of biomedical research—cervical cancer cells taken from a patient named Henrietta Lacks by Johns Hopkins researchers without her knowledge in the 1950s. Hers became the first cell line researchers figured out how to keep alive perpetually. Lacks herself succumbed to the cancer soon after her cells were taken. The cells have outlived her by more than fifty years.

With all the materials in place, the pair huddled around the table and smeared white blood cells on the plates where the cancer cells grew, sealed them in the Ziplocs, placed them under the microscope, set the camera to shoot time-lapse video, and went to bed.

The pictures tell the story. A dozen of Gentry's cells, called granulocytes, swarm the cervical cancer cell, which dwarfs its attackers in size. Like piranhas smelling blood in the water, they surround their larger prey and batter the cancer cell until it bursts.

"We were lucky, because on the first day we tried it we saw that my immune cells didn't do anything, [but] Eri's immune cells blew up a cancer cell!" Schloendorn said.

The kitchen table experiment was only the start of Schloendorn and Gentry's DIY research, but its results laid bare the crucial questions. How did Gentry's cells apparently destroy the cancer cell, and more important, why did they target it? Also, why did Schloendorn's cells not do the same thing? Researchers have made some progress toward a better understanding of the mechanics of the immune system's cancer-fighting powers, but the central medical mystery endures: Why do some people get cancer while others do not?

Schloendorn and Gentry's desire specifically to probe the immune system for answers was driven by several findings. Cancer or lack of cancer appears to run in families. A person's chance of getting cancer increases with age but levels off in very old age, which could mean that some people not only do not but cannot get the disease. Schloendorn also came across a 1957 study appalling in its methods but intriguing in its results. An account from *Time* written the year of the study describes what happened:

> On wooden benches in the well-guarded recreation hall of the Ohio Penitentiary at Columbus sat 53 convicts—killers in for life, bank robbers, embezzlers, check forgers. Some wore the white jacket and trousers of hospital attendants (duty for which they had volunteered in the prison); others, fresh from work gangs, wore blue dungarees. As a man's name was called he walked upstairs to a room equipped as an emergency surgery, sat down and proffered a bare forearm. Dr. Chester M. Southam of Manhattan's Sloan-Kettering Institute then proceeded to inject live cancer cells.

Southam wanted to find out whether a healthy person's immune system would reject the transplanted cancer cells the same way the body rejects transplanted organs and other tissues. The *Time* article goes on to describe the blobs of fluid under each man's arm containing millions of cancer cells. Some of the prisoners' arms swelled and turned red and tender—the immune system at work. Others felt little at all. After two weeks a surgeon cut chunks of flesh from the prisoners where Southam had injected the cancer. Back at Sloan-Kettering, he found that the cancer had disappeared or was very nearly gone for everyone who had been injected.

Southam's experiment demonstrated the immune system's power against cancer cells. But which kind of immunity was at work?

The body's immune system has two branches: the adaptive and the innate. The innate immune system carries in its operating instructions what amounts to a kind of pattern recognition software that determines which foreign invaders the body will recognize on contact and seek to destroy. Over the history of human existence the body has become wired to block automatically a host of familiar pathogens, from measles, mumps, cold, and flu to strep, staph, various fungi, and salmonella.

The adaptive immune system, on the other hand, can learn to recognize germs not already encoded in the innate immune system. Relying on such evolutionary marvels as T-cells, the adaptive immune system exhibits a complex biochemical "intelligence" that can take the measure of an unwelcome microbe and customize a response against it. And once the adaptive immune system learns, it does not forget. Prior to the development of the proper vaccine, patients only suffered through measles once. If they survived, their bodies would easily fight off any subsequent infection by the virus, thanks to the continuing presence of antibodies developed in response to it. Vaccines work according to the same principle: by tricking the adaptive immune system. Edward Jenner's smallpox vaccine used the harmless cowpox virus to provoke the adaptive immune system into developing antibodies that, because smallpox and cowpox germs shared enough of a biochemical resemblance, would fight off both. Other vaccines, such as some polio and flu shots, expose the body to dead versions of viruses that the adaptive immune system will nevertheless scan and learn to shut out.

Of the two branches, researchers have found that the adaptive immune system made more intuitive sense as the place to seek cancer resistance. Since cancer patients do not appear to have an innate immune response when afflicted with cancer themselves, maybe antibodies could be custom designed to fight the disease. Perhaps, as with vaccines against viruses, the body could be trained.

In the spring of 2010 the U.S. Food and Drug Administration approved Provenge, a vaccine developed by Dendreon Corp. to treat advanced prostate cancer. Hailed as the first approved therapeutic cancer vaccine, Provenge has made its developers millions even though the medicine itself mostly consists of the patient's own cells. To create the doses, immune cells are obtained from the patient's blood and exposed to a protein found in most prostate cancers to stimulate an immune response. The cells are then injected back into the patient, carrying what amounts to a wanted poster for prostate cancer. The immune cell posse targets prostate cancer that has spread throughout the body. Because they specifically target only cancer cells, Provenge leaves healthy cells alone, limiting the hideous side effects usually associated with chemotherapy. Clinical trials found that Provenge extended the life of men with the disease by more than four months.

For Schloendorn and Gentry, four months did not impress. What's more, they believe that the results from other studies of fighting cancer using the adaptive immune system showed limited potential. They believed that the adaptive immune system may be too clever, that its sophisticated mechanisms compared to the brute force approach of the innate immune system made it more vulnerable to being fooled. The pair tracked down several studies showing the ways cancer appears to shut down the adaptive immune response—by persuading would-be attackers that the tumor cells belong. So they decided to focus their search for a cure on the body's inbuilt response. Perhaps the more primitive part of the immune system held more promise for an actual cure.

With video evidence in hand, Schloendorn said he was able to get a modest investment—enough to move the lab out of the kitchen and into the garage.

Once there, Schloendorn took full advantage of the lab robot he was able to acquire with the new money. He ran more tests like the one he and Gentry had first done at the kitchen table. He shot video of more granulocytes—the workhorses of the innate immune system—clobbering different kinds of cancer cells.

The pair was excited about their work. But they also had an anxiety unique to do-it-yourself drug development. To help fund their research and give others the chance to do their own work in their own way, Gentry and Schloendorn founded a nonprofit called Livly. Most charities raising money to fight cancer seek publicity as a core function of what they do. Breast cancer charities advertise fund-raising walks on subway trains. Lance Armstrong puts yellow rubber bracelets on the arms of millions. Meanwhile, Gentry and Schloendorn turned down interview requests and kept a low profile. They had not cured cancer yet, but they felt certain about one thing: They probably should not tell anybody about how they were trying.

The block where they built what was likely the San Francisco Bay area's most well-appointed semisecret garage cancer research lab is lined with modest single-family homes. It is easily accessible from the freeway. In-N-Out Burger is one of many casual dining choices along the commercial strip a few streets away. It is a clean, quiet, and utterly typical Silicon Valley suburb.

The setup hidden behind that exterior, however, was hardly typical, even for a Silicon Valley home, and Gentry and Schloendorn knew it. Instead of router racks and fiber-optic cable, they had outfitted their garage with microscopes, cell incubators, and the liquid-handling robot. The Livly Web site proudly described the lab's arsenal of gear, such as a subzero freezer (literally: negative 20°C) and a Robbins Scientific blood mixer ("Need a really gentle shake? The Robbins could rock a baby to sleep").

Despite the lofty idealism of their goal, Gentry and Schloendorn feared their lab could get them into trouble. Because so little precedent existed for what they were doing, they could not know for sure what kind of reaction to expect, legal or otherwise.

"If you can have a conversation with everybody who might be scared of that sort of thing, you could, with logic and truth, overcome their objections, but unfortunately that's not possible," Gentry told me. "We were afraid of having a large negative reaction that might cause us to have to shut down the research. We're not doing the research just for fun or to have something in our garage. It's our key motivator. If that were shut down, what were we going to do?"

Over time they began to take tentative steps toward letting more light shine on their effort. Gentry began organizing meetups for Bay Area residents interested in DIYbio. Some of the early talks included presentations on gene design, algae production, and biotech in developing countries. Their enthusiasm and their gear, along with a certain amount of salesmanship Gentry picked up while working in Apple's marketing department in college, made their home a hub. Their couch hosted other outsider bioengineers eager to work together. Not everyone wanted to splice genes or create the next great wonder drug. But everyone was joined by the idea that hacking should not be limited to computers.

As interest in DIYbio grew, investors looking for the next big thing began sniffing around some of the meetings. In the spring of 2010, Schloendorn gave a presentation on his own work. After the talk, he said, investors showered him with dollars. Like so many others in Silicon Valley before him, he was ready to move out of the garage. Except he wasn't taking a new piece of software with him. He was taking petri dishes.

In the San Francisco Bay area, the Maker Faire is a geek high holiday. The annual event brings together Burning Man fire-sculpture builders with executives from Tesla Motors pushing their $100,000 electric roadster, steampunk bike builders with high school science fair prodigies showing off their newest robots. The centerpiece of the 2010 Faire was a five-story, five-ton retro future rocket ship out of the pages of a 1930s comic book. The rocket did not really fly, but that isn't really the point of the Maker Faire. The rocket stood for the countless hours and passionate exactitude of all geeks who willfully create something out of nothing, social approval or even basic utility be damned. You do it because that is what you do.

The previous year, biohackers had one booth at the Maker Faire.

This year, they had four. Together in the main expo hall, they showed the giddiness of kids who were finally allowed to sit at the grown-ups' table. They had their pitches down. They were being taken seriously, or at least as more than a pure novelty. By the standards of the Bay Area's welcoming yet rigorous DIY subculture, they had made it.

"Being here is like Disneyland times ten," said Gentry. This year Gentry was on center stage, literally, speaking from the Faire's hub of public conversation about what it means to make. For months prior, she had avoided drawing attention to her and Schloendorn's lab. Now she was ready to sell her vision to anyone who would listen.

"Biotech is for everybody," she told the audience. "You can make a difference if you have that right intention, if you have the passion that drives you. If you have that community and access to tools, the world is your oyster."

The hacker space is a particularly Bay Area tradition. Spots like Noisebridge in San Francisco and Hacker Dojo in Mountain View are equal parts oversized dorm room, secret kids' clubhouse, and zealous start-up incubator. They bristle with the intellectual energy of geeks whose enthusiasms have overflowed their ICQ chat rooms and need the embrace of actual human contact. Coding still reigns as the main preoccupation of most hacker spaces, though shelves stuffed with circuit boards, copper wire, and soldering irons speaks to an equal passion for electrical engineering.

Even in these spaces, however, biotech has not gained interest as more than a curiosity. Tinkering was just not done with things that were wet. Now Gentry was ready to change that. Tito Jankowski, Joseph Jackson, and she had joined forces in an effort to open BioCurious, the first hacker space for biotech. The space would be anchored by some of the cast-off gear she and Schloendorn had bought off eBay, Craigslist, and big biotech companies shedding old gear at auction.

The vision for BioCurious extends beyond access to the gear, however. Altering the basic chemistry of life also requires at least a basic working knowledge of the structure of cells, the design and function of DNA, and the way genes are translated into proteins and proteins into the stuff of everyday existence. Gentry believes all those things are eminently teachable to nonspecialists. For Gentry especially, the joining together of like minds was at least equal in importance to the work she and Schloendorn were trying to do in the garage lab.

"I like putting people together who can collaborate and help one another create something big," Gentry told the audience. "At a place like a hacker space, you can come together. You can ask people for help. You can feel comfortable, you can have fun, which is I how I've found ideas really flow. And then, once you form the ideas, you have a full set of biotech equipment where you can actually work those out and invest in your dreams."

Gentry expresses her idealism smoothly, like a pitch she has practiced and polished until she can recite it effortlessly. But her confidence is different from that of the typical start-up entrepreneur, where earth-saving aspirations often prettify a plain old plea for more cash. BioCurious is a nonprofit, and fundamental to its philosophy is that anything invented there belongs to whoever invented it. If someone cures cancer at BioCurious, the lab will make nothing off of licensing fees, one of the largest sources for revenue for university labs in the past thirty years. Gentry says that's okay.

"The kind of person I have in mind is someone like John, someone who is stuck in a place where they know how to innovate, how to solve problems, but they don't have a place to do it," she told me later. "If they have an ability to solve human problems, I want to focus on that, and give them the tools they'll need."

On a warm June night in Silicon Valley, Gentry was getting ready to go to camp. But she had a few things to take care of first. She had found a temporary space to keep the gear for BioCurious, and outside her apartment a U-Haul was waiting to take away the last pieces of the garage lab. A friend with a jerry-rigged DIY steadicam followed her around the apartment shooting a BioCurious promotional video that would go up that night on Kickstarter, the crowdsourced funding site for DIY projects where Tito Jankowski had had so much success with his OpenPCR machine. Inside the garage, an audience of about twenty sat rapt on a Thursday night listening to a two-hour lecture on the U.S. Food and Drug Administration's approval process for new pharmaceuticals and medical devices. In the next room, Schloendorn sat proudly atop a plastic case holding human embryonic stem cells he had ordered online.

Gentry had been tapped to attend Foo Camp, an annual invitation-only gathering hosted by tech publishing mogul Tim O'Reilly at his company's Sebastopol, California, campus. Participants do spend the weekend sleeping in tents, though little sleeping reportedly occurs as the carefully curated group of 250 smart people chatter into the night forging what O'Reilly has called "new synapses in the global brain." Gentry planned to promote BioCurious, specifically a new project that both honored and poked fun at bio celebrity J. Craig Venter—a pioneering entrepreneur and gadfly in both DNA sequencing and synthesis.

At the Maker Faire, Gentry met a boat maker with an interest in biology. The two quickly hatched a plan to mimic Venter's self-styled epic quest to sail around the globe in a yacht to collect and identify new ocean species. Venter's venture fit with his earlier effort to decode the human genome both in its grandiose ambition and because of the novel way he proposed to accomplish it. Rather than collecting an aquarium of big fish and strange sea creatures, Venter and his crew would gather microbes too small for the human eye to see. He would then use his own sequencing technology to take the genetic fingerprints of the species he collected to determine whether they were actually new to science. Since 2003, Venter and crews under his command have sailed the world on multiple voyages he says are inspired by Darwin.

Gentry's excursions are closer to home, in keeping with the DIY penchant for the domestic. She has not ventured beyond the waters of San Francisco Bay. Yet her voyages have at least as much in common with the spirit of Darwin as anything Venter has tried.

Gentry extracted the DNA of the samples she last collected while on the bay using simple gear at a friend's house. Since few desktop sequencers have hit the market at prices anyone outside a big research lab can afford, she said she planned to outsource her sequencing to a mail-order company for $100 a sample. She winced at the price—far too high for any respectable DIY project. But she said the science was worth it.

"You're discovering new species no one has ever seen before. How can you put a price on that?"

Later I met Schloendorn a few towns over, in Menlo Park, home to Sandhill Road, which is to tech entrepreneurship what Wall Street is to high finance. Along this leafy thoroughfare just west of Stanford University, some of the world's most storied venture capital firms hatched deals that have transformed the global economy. Across town in Menlo Park's version of the wrong side of the tracks sat one of the most boring buildings ever built. Glass doors like the kind that front every convenience store broke up the monotonous white two-story facade. No colorful signs told what hid behind these walls; street numbers in a slightly tacky serif font vaguely suggested the 1980s.

Behind one of these doors, Schloendorn had a new garage. The ceiling was at least twice as high as the garage back at the Mountain View apartment, and the space had a second story where he had thrown down a bare mattress and installed a worn wooden table. He also had something else the first garage lacked: employees. In the back of the main space a lanky, unshaven young man in a T-shirt was having halting success getting a computer to move a robotic arm back and forth. Several other robotic arms lined nearby tables. The more of these robots he could set to filling plates and running assays, the more Schloendorn and his staff could sit back and let their minds rather than their hands do the work.

Gone were the Ziploc bags and dry ice; in their places a temperature- and humidity-controlled incubator was being fed by a carbon dioxide tank. Inside, several trays of cancer cells awaited possible destruction. Schloendorn did not plan to rely on his own blood to fuel his start-up's research anymore. Instead, he expected to use human embryonic stem cells to generate a steady supply of granulocytes to test against various forms of cancer. Even if he turned out to be completely wrong about the innate immune system's ability to fight the disease, he said, he would at least have a handy little business manufacturing white blood cells. Biotechs can always use more of those.

The entire lab setup ran about $30,000, Schloendorn told me—not cheap, but not glamorously expensive, either. He said the half million dollars will last his new company about a year, which meant he needed to show some kind of result within six months to dangle in front of other investors if he wanted to continue for another year. In the parlance of mainstream drug development, these intermediate results are called milestones, the basic currency of institutional biomedical research. Schloendorn was clearly tickled by his return to the supposed dark side—this was not an open-source project—but he doesn't much care how the work gets done as long as he gets to do it.

"Most institutions tend to focus on being institutions more than on accomplishing goals. And that's fine if that's what they want to do," Schloendorn said. "I understand that people don't have goals. That's fine. I wish it wasn't necessary for me to have goals. It's just that I'll die if I don't do anything about it."

II

READ/WRITE

Since the first punks took up the DIY banner in the 1970s and 1980s, defiantly rejecting the corporate music industry and the cultural ills for which they felt it stood, the idea of doing it yourself has carried political weight. Woven from threads of anarchism, libertarianism, and Thoreau-style radical self-reliance, the American version of DIY has always exhibited a healthy enthusiasm for entrepreneurial capitalism, even if the early punks starting record labels, nightclubs, and print shops may have shunned the term. In the Silicon Valley version of DIY, that enthusiasm has always been unapologetic and highlights the long, uncomfortable alliance between nerds and punks thrown together in common cause by decades of high school social ostracism.

The particular idealism that ties all DIYers together is an overarching belief in the power of the individual to succeed where institutions corrupt and fail. Autonomy sparks creativity and rewards initiative while bureaucracy hinders change and punishes risk. Open participation breeds innovation while closed hierarchies lead to stagnation and politics over progress.

Yet, as Chris Kelty has observed, biopunks would be nowhere without the Man. Most of the technologies that have inspired biopunk and make the work possible emerged over the last forty years out of university, business, and government labs. In the 1970s, recombinant DNA was developed at Stanford and UCSF, still globally prominent hubs of biotech research. University researchers figured out around the same time the first effective methods for creating monoclonal antibodies, cells customized to target specific disease-causing cells and crucial to the success of bestselling biotech drugs. In the early 1980s, Berkeley-trained researcher Kary Mullis developed the PCR technique at Cetus, one of the world's first and at the time best financed biotech companies. More recently, the Human Genome Project shines as the brightest example of an achievement that only Big Science seemingly could have pulled off. Billions of dollars. Dozens of labs. More than a decade of work. Hundreds of federally funded researchers. And in the end, a singular accomplishment: a complete map of our genetic selves.

Today's life-science DIYers depend on the latest information technology. Personal computers and mobile devices have become ubiquitous, as has the Internet, enabling unprecedented digital autonomy for individuals. At the same time, each of these inventions exists as a kind of epitome of the entire sociopolitical economy of modern capitalism. The Internet, after all, began as a military project.

So is it disingenuous to build a subculture atop such a massive institutional foundation and call it DIY?

Maybe not. Thoreau never really became self-sufficient out there in his cabin by Walden Pond. Yet his account of the experience in *Walden* transformed notions of American identity and inspired generations to rethink their relationships to society, nature, and themselves. In the same way, the real significance of DIY biotechnologists might lie not in any particular technological achievement but in the provocative questions they raise. They may never cure cancer. Yet their idealism and critique of the scientific establishment could make a lasting impact. In reality, American DIY movements have rarely been about dropping out of society but about clever reimaginings of social norms. Biopunks do not build their own lab tools to maintain some kind of purist separation from the existing system. They want to force the conversation about how that system works and who it serves.

Kelty contends that the provocative name of his conference, "Outlaw Biology?", forces a confrontation with the issue of why do-it-yourself biotech makes anyone anxious in the first place. Ultimately, he says the answer has less to do with some sensational biotech apocalypse and more about the destabilization of how science is done: "Who sets the agenda? Who really innovates? What can individuals actually do amongst the massive juggernaut that is Big Bio and Big Pharma today? Who gets to tinker with what?"

Today the answer to the first part of that last question is "more people than ever before." The development of two intertwined technologies over the past ten years has forever changed how we will relate to our biology and who gets to be involved. Reading and writing DNA have become easier than ever before, and in ways that blur the lines between Big Science and all-access. Do-it-yourself bleeds into direct-to-consumer gene scans, and teenagers become the pioneers of building machines made of genes. Genetic engineering could someday become as easy as booting up a laptop if the pioneers of synthetic biology succeed. Cheap DNA sequencers and cheaper digital storage mean each of us could soon carry a scan of our entire individual genome on our smartphone. As reading and writing DNA becomes more and more like processing bits and bytes, the closer genetics comes to being a part of everyday life. As that happens, we could all find ourselves becoming DIY biologists.

CHAPTER 11

Reading

In the 1980s, he stalked the streets of one of Los Angeles's poorest neighborhoods strangling and gunning down women—seven in all before the shooting stopped, at least for a while. Starting in 2002, the murders began again. The press called him the Grim Sleeper because of the long gap between killing sprees. Investigators had leads, but the detectives on the streets never nailed their man.

Meanwhile, hundreds of miles north in Richmond, California, another impoverished city that knows its share of killing, government gene jockeys in a lab were running a routine check when something unusual popped up. In a practice known as familial DNA search, the state-run lab was comparing the DNA of a man recently convicted of a felony weapons charge to DNA profiles stored in a California crime database. Against the objections of privacy advocates, California law enforcement officials in 2008 began searching for genetic similarities between convicts' DNA and stored DNA evidence from unsolved crimes. Instead of just looking for direct matches, investigators also started to seek similar Y chromosomes, which are shared among male relatives. Matching Ys might link not just convicts but sons, uncles, brothers, and nephews to crime scenes.

In this case, thirty-one-year-old Christopher Franklin's DNA partially matched samples recovered at several Grim Sleeper crime scenes. Suspicion turned to Franklin's father, fifty-seven-year-old Lonnie Franklin, Jr. Agents staked him out, swiped a pizza crust he had thrown away as they followed him, and from it lifted a current DNA sample. The sample matched. Investigators said Lonnie Franklin Jr. was the Grim Sleeper. He has pleaded not guilty.

The use of familial DNA search to catch a serial killer was one of the most Hollywood-ready uses of comparative genomics to date. But law enforcement hardly has a monopoly on the technology used to track him down. For as little as $149, anyone with an Internet connection can order the same test used to catch Franklin from Ancestry.com to track down "genetic cousins." The site claims customers can use the results to trace paternal ancestors dating as far back as one hundred thousand years.

Once the exclusive realm of high-end science and homicide investigations, DNA has become part of the everyday. The grandly expensive and decade-long Human Genome Project has in less than a decade yielded to fast, cheap sequencers that can decode anyone's DNA for the price of a cheap used car. (Many in the field believe that within ten years the cost of getting an entire human genome sequenced will close in on zero.) Societies have only just started to digest the implications of this dramatic change in what we can know about our biological selves. Yet a growing group of not just biopunks but regular consumers have decided what unbridled access to DNA means to them. They want to know more—about themselves, about others, about every living thing—and life not yet imagined. And they do not want to wait for anyone else to tell them what they can and cannot know.

As the opportunity to know our genome's three billion letters falls within reach of middle-class consumers, science's understanding of the words those letters spell is expanding with similar speed. Rapid DNA sequencing has allowed scientists to undertake broad studies of genetic differences among individuals that reveal what specific genes actually do and what kinds of risks to health they pose. These studies have yielded insights into obesity, mental illness, life-threatening diseases, everyday behavior, and more. Even now, few aspects of our selves are escaping genetic scrutiny. While few geneticists would argue that nature always trumps nurture, most would assert that genes are always involved in the equation.

The plummeting price for sequencing DNA has long fueled predictions that everyone's medical record will soon include a copy of his or her genome. Businesses are staking millions on so-called personalized medicine, the promise that decoding our genes will lead to better diagnosis, treatment, and prevention of disease. For biohackers, cheap DNA scanning means something else.

"Once it becomes five dollars cheap to sequence environmental DNA, people are just going to sequence everything," Mackenzie Cowell told me.

Right now, five dollars does not get you a long enough reading along a strand of DNA to learn much of anything useful. The sample you are scanning must include a sea of copies of the same gene sequence, and the sample must be pure. Too many bits and pieces of other genes will distort the scan and lead to an inaccurate reading.

With third-generation sequencing technology, those requirements are loosening, and the price is dropping. As with software that tries to predict what you are going to type or where you want to focus the camera, the latest sequencers use sophisticated algorithms to filter out the genetic noise. As that technology improves, so does the capacity for off-the-cuff sequencing. Biohackers who want to know what microorganisms they have tracked into the lab will be able to swab the bottoms of their shoes, stick the sample in a vial, and mail it off. For a few bucks a sequence they will get an e-mail back showing what they have found, letter by letter.

While do-it-yourself biotech remains a small subculture, direct-to-consumer genomics is bringing the age of genetics into homes across the United States and around the world. More than any other recent phenomenon, DNA-scanning start-ups have served as an indicator that genetic information will no longer stay hidden inside labs at elite universities and hospital clinics devoted to rare disorders. While we are all creatures of our genes, we may all soon participate much more directly in decisions informed by what we know about our genetic profiles. In a sense, we will all become DNA tinkerers, making decisions about our lives and lifestyles in an effort to tweak our futures based on what our genes appear to have in store for us.

At the beginning, the press fawned. And how could they not? When Anne Wojcicki and Linda Avey launched 23andMe in 2007, only a few science celebrities had had the three billion letters in their DNA sequenced. Now Avery and Wojcicki—the wife of Google cofounder Sergey Brin—were telling consumers they too could unlock the secrets of their DNA for a mere $999 and a vial of spit.

To broaden the appeal to those not immediately seduced by the science, the company began hosting celebrity spit parties. A Talk of the Town piece in the September 22, 2008, issue of *The New Yorker* captured the mood best:

> Avey and Wojcicki were joined by Wendi Murdoch, Rupert's wife, who had already taken her DNA test. "I did it," she said. "My children did it. Rupert's mom did it. She's ninety-nine years old!" She logged on to a laptop near the cheese table and pulled up a profile of her inherited traits, such as alcohol-flush reaction. "My daughter has that, too," she said. "But Rupert doesn't. He can drink, but he won't get red."

Celebrities were spitting, but you did not need the fortune of a Rupert Murdoch anymore to peer deep into your genetic self. Seemingly overnight, 23andMe had made available for the cost of a laptop computer or a month's rent what had recently teetered on the mind-boggling edge of biological knowledge.

The price and speed of sequencing DNA had changed more quickly than nearly anyone had predicted. Companies like 23andMe could scan a customer's DNA for a minuscule fraction of what the first full genome scan cost U.S. taxpayers. More important than the scan of individual customers' genes, however, was what cheap sequencing promised for biomedical research. Now scientists could scan DNA for variations they hoped could explain the genetic roots of any human trait or medical condition. So far, while the sequencing has become easy, the biology remains hard. But science appeared to have taken another step toward connecting who we are and what afflicts us to our DNA.

Such experiments are known as genomewide association studies. The premise is simple. Take any physical trait—for example, the tendency to turn red when drinking alcohol. A scientist would find two groups of people—one that turned red and one that did not. The DNA of each group would be scanned—not all three billion letters, but in areas where researchers had some reason to believe the trait and the gene would be linked. Time and again, statistical analysis would show some letters differing consistently between the groups along a certain segment of DNA. For instance, in the case of alcohol flush, having AA along the ALDH2 gene is associated with extreme flushing, an AG with moderate flushing, and a GG with no flushing. Such correlations led researchers to believe that they had found a way to home in on the genetic source code of nearly any aspect of the human species.

Since those heady days of 2007, researchers have discovered that variations in DNA offer only the first clue toward understanding the complex system linking the genome to specific traits. Until they gain better insight into the basic biology of many diseases, the link between specific conditions and particular DNA sequences may remain little more than interesting statistical correlations. Still, the mounting data from genomewide association studies offer the first inkling of insight into one of the greatest mysteries of the self. The first 23andMe gene scans covered about five hundred thousand sites along the genome where scientists had identified promising variations, known as single nucleotide polymorphisms, or SNPs (pronounced "snips"). For the most widely researched SNPs, customers can log into the 23andMe Web site to learn which traits they possess or their risk of developing a particular disease. This last feature almost immediately became a source of controversy that has only grown as the company and its services have become more widely known.

Unfortunately for 23andMe, the media's fascination with popular science has an even shorter half-life than its celebrity-of-the-month obsessions. As the glowing accounts faded, the challenge of running a viable business set in. The trouble began shortly after the company slashed the price of its test to $399—at the time the price of a good iPod—in a move pitched as a way to bring personal genetic information to an even wider group of consumers. Speculation ensued that the company's business model was out of whack, and that the price of gene sequencing was getting so cheap that 23andMe would soon have little to sell that couldn't be had nearly for free. The company endured layoffs. Linda Avey departed. The demise of a competing service heralded a new round of press, this time musing on the possible death of an industry still in its infancy.

Then the government started asking questions.

In July 2010, the week before politicians and regulators would force personal genomics companies to answer for themselves before Congress, 23andMe hosted a policy forum at a hotel south of San Francisco, near the airport. The assembled scientists, bioethicists, entrepreneurs, lawyers, pundits, and journalists had a range of opinions on the usefulness and risks of mass-market genomics, but few voiced any doubt that such information would become a central feature of twenty-first-century medical practice.

Wojcicki struck a placating tone before the largely sympathetic audience. She pointed to her company's own effort to regulate the fledgling direct-to-consumer genetic testing industry. She said 23andMe helped craft California's Senate Bill 482 to provide "meaningful oversight" of companies like hers. The bill would require companies to hire experts to set credible standards for interpreting customers' genetic information. The standards would have to be public, and each company would have to organize an external physician advisory board. Under the law, no company could advise customers about medical treatments.

The American Civil Liberties Union objected to the bill, saying it did too little to protect patient privacy. Other groups claimed the measure, which as of mid-2010 had stalled in committee, would only add more confusion to an already muddy regulatory situation.

As Wojcicki spoke, the situation for 23andMe was about to get muddier. The U.S. Food and Drug Administration was preparing to grill 23andMe and its major competitors in just a week's time. Congress had called its own hearing for a few days after the FDA interrogation.

Federal regulators and politicians had mostly ignored companies like 23andMe while the services remained online only. All that changed in May 2010 when a competitor, San Diego–based Pathway Genomics, announced plans to begin selling its genetic testing kits at thousands of Walgreens stores across the United States. The test was to be no different than what the company already offered online, and making the product available in the drugstore was really little more than a marketing ploy. The box on the shelf would contain the same vial the company was already mailing to online customers to collect their saliva, which the spitters would then send back to the company for testing. But the media attention spurred by Pathway's decision to make direct-to-consumer genomics much more direct caused a strong shift in the political winds.

Since their launches, direct-to-consumer genetic testing companies had argued that regulators should not classify their products as medical devices because they merely provide information, not diagnoses. On its Web site and in its marketing materials, 23andMe clearly states that customers should never make medical decisions based on their test results but should always consult with a doctor. That did not placate consumer advocates, who pointed out that very few physicians had any training whatsoever in genomic medicine. Moreover, research into most links between SNPs and medical conditions was preliminary at best.

States had also been wary. Public health officials in California and

New York had sent cease and desist letters to 23andMe and others in June 2008 ordering them to stop selling their products until they verified that their labs met state and federal standards and that licensed physicians had ordered the tests for their patients. After a flurry of media attention, California regulators did an about-face and issued both 23andMe and competitor Navigenics licenses to operate after assuring themselves that the companies' test results were grounded in the scientific literature, and that doctors were reviewing customers' orders.

Just days before Pathway's kit was to hit shelves, the FDA sent the company an enforcement letter saying the test qualified as a medical device but had not received FDA approval. Walgreens quickly scrapped its plan to sell it. About a month later, several Pathway competitors received similar letters, including 23andMe.

For at least two years after direct-to-consumer DNA scans went on sale, federal regulators and politicians sat idle as bioethicists and public health officials griped about the risks. As soon as the industry that had been serving a niche market of nerds tried to go mass market, government decided the gripes suddenly mattered.

House Energy and Commerce Committee chairman Henry Waxman called the congressional hearing. In his opening statement he chided the four most prominent direct-to-consumer companies for marketing their genetic tests as guides to better health despite "no widely accepted consensus linking genetic markers to specific illnesses." The Santa Monica Democrat said that he applauded the FDA's crackdown on the services. During a question-and-answer session, Michigan Democrat Bart Stupak referred to personal genomic scans as "snake oil."

Yet the coup de grâce came not from the FDA, Waxman, or Stupak but from the Government Accountability Office, Congress's investigative wing. At the hearing the GAO revealed that it had conducted a yearlong investigation of fifteen companies. Their report's title tidily summed up the GAO's conclusions: "Direct-to-Consumer Genetic Tests: Misleading Test Results Are Further Complicated by Deceptive Marketing and Other Questionable Practices."

The most sensational moment in the hearing came when chief GAO investigator Gregory Kutz played a YouTube video of recorded phone calls between representatives of unnamed companies and undercover operatives acting as patients. The "patients" had called the companies to follow up after receiving their test results. In one call a patient was told she was at high risk for getting breast cancer, even though she had not been tested for mutations in the gene most closely tied to the disease. In another a patient was told that therapies existed to repair damaged DNA, a claim with no scientific basis. "The genes are considered now not to be the source of our biology. They're a symptom," the rep told the patient.

In the report itself, the GAO criticized testers for making risk predictions at odds with patients' family histories or actual medical conditions. One of the most oft-cited examples in news accounts of the hearings was a patient with a pacemaker who was told he was at below-average risk for the very heart condition he had been diagnosed with years earlier. The GAO used this and similar experiences of other patients to underscore an unnamed expert's claim in the report that "the most accurate way for these companies to predict disease risks would be for them to charge consumers $500 for DNA and family medical history information, throw out the DNA, and then make predictions based solely on the family history information."

Supporters of direct-to-consumer genetic testing services saw the hearings as a dark day that foreshadowed future regulations that could kill the industry before it had even left the nest. (An FDA representative at the hearing promised that the agency was exploring its options.) Critics of the GAO report seized on the pacemaker example as evidence of at best a misunderstanding of the meaning of risk versus diagnosis. They also said the report failed to distinguish between blatant charlatans and companies making a good faith effort to follow best scientific practices, such as 23andMe, San Francisco Bay Area–based Navigenics, Iceland's deCODEme, and Pathway.

"Now that the report is public and we have had a chance to review it, we are troubled and find the report is deeply flawed," 23andMe wrote on its blog in response to the GAO investigation. "This report raises questions, but leads to few conclusions because of its unscientific nature. . . . We are confident in our service's accuracy and reliability."

During the back-and-forth, the hearing never turned seriously to the underlying philosophical issue at stake for biopunks. When the FDA regulates a blood test that determines whether a patient has, for example, hepatitis or HIV, the agency is evaluating a device that provides a yes or no answer. The test accurately tells whether a patient has the infection or not. Quality control is black and white.

Prediction of risk affords no such certainty. Probabilities can grow more statistically assured. But at the current level of understanding, the predictions of personal genomics tests will always come out gray. (Personal genomics tests should not be confused with tests for congenital disorders like Huntington's disease, which are caused by specific and well-characterized mutations.) Waxman, Stupak, the FDA, and the GAO worry that gene-scan recipients may take their test results more seriously than the science warrants and potentially make medical decisions based on bad information. That concern does not persuade gene geeks that such tests need regulating. They would argue that federal authorities would do better to try to improve the average citizen's understanding of probability. In the meantime, biopunks ask, should one person's intellectual shortcomings infringe on another's right to information about themselves?

The controversy over medical information provided by 23andMe and other direct-to-consumer DNA-scanning services hinges on the twin questions of accessibility and accountability.

Public health officials and ethicists warned early on that consumers could be misled by the very preliminary findings used by 23andMe and others to rate future health risks.

"We just don't know how people will use the information," Dr. Jinger Hoop, a professor of psychiatric genetics and medical ethics at the Medical College of Wisconsin in Milwaukee, told me in 2008. "We don't know whether it will be helpful to them in the long run."

Some critics worried that the information could in fact do harm. At the 23andMe policy forum, medical ethicist Amy McGuire of the Baylor College of Medicine recalled the episode of the *Oprah Winfrey Show* featuring 23andMe. McGuire described how during the show, Winfrey's in-house physician, Dr. Mehmet Oz, mentioned that his 23andMe scan showed he had a low genetic risk for prostate cancer. According to McGuire, Oz said that the results meant he would not have to subject himself to the unpleasant regular prostate screenings recommended to men over fifty.

McGuire said she began shouting at the TV because she believed Oz—an accomplished surgeon—knew better than to believe the 23andMe test offered anything like conclusive evidence upon which to make such a serious medical decision. She was willing to believe he was being facetious but also that less sophisticated viewers might not pick up on the joke. She and others continue to worry that without adequate counseling and consultation, consumers could be led to make choices like this based on their results even without the apparent approval of a celebrity doctor.

Supporters counter that everyone has a right to their own genetic information, and that restrictions on services like 23andMe would be government overreach. What right do regulators have, the sentiment goes, to tell people what they can and cannot know about themselves?

"I have little patience for the argument that we need doctors as gatekeepers of our genetic information," Thomas Goetz, executive editor at *Wired* and an early champion of personal genomics services, wrote on his blog.

> This isn't a drug, and this isn't a device—it's information about ourselves, as ordinary as our hair color or our waist size or our blood pressure—all things that we can measure and consider without a doctor's permission. . . . To

me, getting access to this information is a civil rights issue. It's our data.

Raymond McCauley lives in a quiet subdivision just off the freeway in Mountain View, California. His house consists mainly of one great vaulting room, more loft than suburban house, that had been taken over by his three-year-old twins. One of the boys ran over when I arrived and asked to play.

"After Daddy does some work, we can either go on an ice cream adventure or a treasure hunt. Which would you like?" McCauley asked. (The boy picked the treasure hunt.) McCauley told me his son's name was Harlan, after the prolific science-fiction writer Harlan Ellison.

McCauley took me on a quick tour of the house, which he and his partner, Kristina, had renovated themselves. In the master bath, the ceiling followed the slanted roofline up over a shower the size of a studio apartment and twice as high. Around the perimeter of the entertainment center, below a massive wall-mounted flat-screen TV, the parents had erected a colorful kid fence.

"When we first moved in, we thought it was going to be all elegant and modern," McCauley told me, affectionately eyeing his family curled up on the couch. McCauley is tall and broad, with a head of thick white hair trimmed short and sticking straight up above black-rimmed glasses. His complexion has a hint of the paleness native to those who spend most of their lives in front of a computer. But he has the can-do energy and still a bit of the drawl of his native Texas, where he spent most of the first three decades of his life.

Another renovation to the house was a steep, narrow staircase with steps more like rungs that lead up to McCauley's work cave. The stairs were blocked by another kid fence, a safety measure that divides one half of McCauley's life from the other. As he put it, the work he does in the cave has to be so interesting that he can justify to himself spending time upstairs instead of playing with his kids.

McCauley has a hacker's typically restless background: unable to zero in on a single knowledge niche, constantly pulled to new jobs, new places, and new specialties by what he described as his raw curiosity about how things work.

McCauley went to Texas A&M in the 1980s to study computers. "That was sort of the digital age right there, the belle époque. And the possibilities were really great," he told me. "But one of the things that always frustrated me was, if you were someone who was just sitting on one side of the computer, you didn't really understand how it worked. You didn't know what the limitations were or the best way to do things. It became important to me to dig inside and open it up."

He double majored in computer science and electrical engineering but also grew interested in genetics, thanks to a girlfriend who was studying to be a veterinarian. ("A triple major seemed like a really bad choice.")

After graduation he worked for a time as a computer programmer at NASA in Houston. He then went to work for the state, an environment he described as a standard bureaucracy in which people came in late and left early, but made sure they got their twenty-five years. McCauley told me he was in a meeting where coworkers were typically unenthused about solving the problem on the table. "Somebody said, 'C'mon, it's not like we're curing cancer.' I was, like, 'Yeah, we're really not.'"

He went back to Texas A&M to study biochemistry and biophysics in hopes of doing something more meaningful. At the time, the Human Genome Project was getting under way, and he quickly saw how he could fuse his skills and interests. He inferred correctly that the new ocean of data unleashed not just by the project itself but the new technologies used to generate that data would change how biology was done. No longer would research be dominated solely by what he saw as the feudal system of principal investigators instructing indentured graduate students, who were locked up in their isolated labs, never communicating or examining how biology's various pieces fit together.

"That was going to change," McCauley says. "People were going to be looking at systems. And nobody was going to be that good that they could keep it all in their heads."

Bioinformatics is the science of managing and making sense of all the digital data generated by new kinds of biological research, like high-throughput DNA sequencing. By the late 1990s it became clear that researchers could advance biological knowledge without ever leaving their computers. McCauley began studying bioinformatics at Stanford and found his vocation.

During the day, McCauley works as a bioinformaticist for one of the world's top manufacturers of equipment for reading and writing DNA. Bioinformaticists seldom get their hands wet. Instead, they try to organize of the data generated by wet-lab research and turn it into useful information. They sit in front of computers and try to piece together the raw facts of life into meaningful patterns. For McCauley, this means sifting through the DNA sequences churned out by his company's machines.

Those machines read DNA by breaking up the genome being studied into more manageable fragments. To each letter in each fragment the machine chemically attaches a microscopic fluorescent tag, a different color for each of the four letters. State-of-the-art optics read the tags, and a computer displays the sequences.

For organisms such as humans that have already been mapped, DNA sequencers can automatically stitch together the entire length of the genome based on an already known pattern. But every species has its own genome. With millions of animal species, plants, fungi, and bacteria and other single-celled organisms, the number of genomes left to map is inexhaustible. And for now, fitting the pieces together for each new organism sequenced still depends on human understanding and intuition.

McCauley's day job extends beyond sequencing new species. Cancers have their own genomic signatures. As McCauley and his colleagues piece together the unique sequences of different types of cancer, they are contributing to the Cancer Genome Atlas, a joint federally funded project to increase understanding of the role genetics plays in the disease. By comparing the DNA of healthy tissue to that of tumors, they hope to give researchers a guide to the good genes that go bad when cancer strikes.

McCauley does not take his work home with him, at least not exactly. His interests sometimes do take him out behind the house to the garage, where he works on a project he will only refer to as his "secret sauce." He won't talk too much about it because, like many inventors, he believes he could have something important in the works, which puts him a little at odds with the biopunk ideal of openness. But like many entrepreneurs in Silicon Valley, he's good at making his idea sound pretty cool.

The machines McCauley's employer and its major competitors make are at the high end of tools available to read DNA. They are still expensive to buy and expensive to run, at least from the average consumer's point of view. At the low end, chips called microarrays let researchers probe for specific short strands of DNA using chemicals tailored to hunt for specific genetic targets. McCauley sees a gap in the middle between those two tools that he wants to fill.

In his garage, McCauley hopes to combine his background in electronics, computer programming, bioinformatics, and nanotechnology to create a device that can read DNA electronically—anywhere. Unlike microarrays, his device would not depend on messy, expensive chemicals to do the job. His DNA probe would rely almost purely on electronics to scan specific DNA samples.

"You could do it really quickly, really cheaply, and reusably," McCauley said. "And I started thinking about all these neat things you could do with electric fields, DNA, and sensing."

He imagines using such a scanner to diagnose diseases in places that lack more sophisticated labs and hospitals by electronically probing for a germ's DNA fingerprint. He envisions a ubiquitous network of tiny biosensors that would sit in your house or attach to a street lamp and register the microbes drifting by. The sensors could serve as an early warning system in the event of a bioterror attack. But McCauley thinks of them more as a microflora "weather report" that could help scientists understand the microscopic ecology of everyday life.

Whether his idea has any chance of working remains to be seen, since he has kept the details secret. But he makes no pretense of expertise. He chose to work in a garage rather than with investors because he did not want to be accountable to anyone else as he pursues an idea he himself calls "weird."

"Instead of going to get millions of dollars in funding and be beholden to people, I'd like to go and do things at my own pace, in my own way and see what I can make happen," McCauley said.

The problem with that approach comes down to the basic difference between a biotech start-up and a venture that never leaves the digital realm.

"At a computer start-up, you can literally do that out of a garage with a couple of laptops. But at a biotech start-up, unless you're just doing computer work, you are making some *thing*. There is a physicality to it," McCauley said. And that physical stuff of a biotech—the gear, the chemicals, the flasks and beakers—"if you really rely on ordering stuff out of the catalog, you're going to max out your credit card pretty quick."

The only other option he sees for a would-be entrepreneur like himself is to go DIY.

"It's a way to jump over that curb of not having the money," he said. "For me the whole DIYbio piece has been about, Hey, there's things I want to do that I can't afford. How can I cut some corners, not from a safety perspective but from a cost perspective?"

While others may go into the garage to stake a claim for their right to research, McCauley describes his own approach as "the entrepreneurial engineer efficiency direction," summed up by a DIY mantra that has taken on new appeal during the recession: "If I can do this for a dollar, can I do it for fifty cents?"

Lucky for McCauley, the DIY moment has arrived. From Etsy to *Make* magazine to home-brewed *kombucha*, going it alone is as cool now as it's ever been. And biohackers like McCauley feel it.

"I think a lot of people who were kind of on that cusp between baby boomers and Gen X remember having their chemistry sets," he said. "It's like, Hey, I learned how to do electronics by taking apart my dad's phone. So where does somebody learn how to do these other things?"

Along with the pleasure of tinkering, McCauley told me that the DIY approach allows him to reconnect with the sense of wonder that drew him to science in the first place.

"You look to be able to capture the fun Eureka! thing. Messing around in a lab in a garage, you can do that."

McCauley's passion for genetic information also led him to direct-to-consumer genomics, the easiest way to find out a few things he wanted to know about himself. He discovered some reasons to worry in the results of his 23andMe test—so much so that he says he lost seventy pounds to improve his odds. But he worried most about the test's finding that he faces a greater than normal risk of developing macular degeneration, a common disease of the retina that damages central vision. The test results claimed he had a 30 percent to 40 percent chance of developing the disorder by the time he reached his fifties.

No treatment exists for the most common version of macular degeneration, nor any certain way to prevent it. Vitamins and antioxidants offer the best hope of keeping the disease from developing, or at least the best hope of feeling a sense of control over the future. Leafy greens contain most of the nutrients recommended.

Folic acid, also known as vitamin B_9, is key to this regimen. But mutations in a gene known as the MTHFR gene—"the motherfucker gene," McCauley said—may impair the body's ability to use the nutrient. The MTHFR gene codes for the production of an enzyme that reacts with folic acid to process certain proteins. According to the National Institutes of Health, researchers have identified at least twenty-four variations along MTHFR in people who have a disease called homocystinuria, in which a lack of that enzyme and others can lead to vision, blood, and bone problems.

The 23andMe test scans for mutations in the MTHFR gene. But the research into exactly what effect the mutations have does not offer any clear conclusions. For someone about to infuse himself with folic acid in the hope of warding off future partial blindness, McCauley did not like this uncertainty. He wanted to know if a specific MTHFR mutation would make a folic acid vitamin regimen pointless, since his body would not be able to use the chemical anyway. Instead of waiting for another peer-reviewed study to come along, he decided to try to figure out the answer himself.

Not that McCauley does not trust the literature. He just does not believe anyone has a strong financial incentive to answer the question that is bugging him. The answers he wants may take a long time to come if he waits for someone else to find them.

"If this is a real thing, and it has a real effect, surely there's some way we could tell it," he said. Specifically, he believed that something in the blood would rise or fall depending on whether the body was processing the folic acid. Staging his own clinical trial felt like the natural next step. So far, science has not developed a decent clinical trial of one. McCauley recruited four other gene hackers who had a different variation in the same location on the MTHFR gene. McCauley was the only one with the variation that predicted the risk of not being able to process the nutrient in vitamin form.

Folic acid functions in the body to spur a reaction that transforms the amino acid homocystine into another amino acid, methionine. Studies have linked a buildup of homocystine in the blood to cardiovascular problems. For people with more serious MTHFR mutations, ones that totally prevent the production of the enzyme that processes folic acid, the consequences of too much homocystine can be even worse. Meanwhile, methionine has several healthful effects, including fat reduction.

To measure the possible effects of his less serious but still possibly adverse gene variation, McCauley's group decided to measure the homocystine levels in their blood. Since the goal of the study was to get results, not to make work, the group did not set up a garage lab and draw the blood themselves. Instead, they went to an online doctor who signed the order for the tests and took their forms to nearby blood-work clinics. They had cheap, accurate results in hand in a few days.

The first blood test measured homocystine levels after the group spent two weeks "washing out," avoiding anything that might contain folic acid. The results were expected to be high, since the homocystine was not being processed well without the nutrient. The four other participants had the expected result, while McCauley's level came in five times lower. Unable to account for the difference, the group next took regular doses of off-the-shelf folic acid vitamins for two weeks. Afterward they found that homocystine levels had gone down for the other members of the group; but McCauley's level tripled, suggesting that the enzyme was not using the folic acid to process it. Finally, they took a different folic acid vitamin marketed specifically to people who otherwise have trouble absorbing the nutrient. For the others, their levels went down further, while McCauley's dropped back to where it had been after the washout. His body appeared to be able to use the better-tailored vitamin.

McCauley did not hesitate to concede that the results do not show anything conclusive. But he still thinks that what he observed was pretty cool. Perhaps his mutation prevents him from using folic acid formulated for standard vitamins but still allows him to take up the B_9 in the other form. At the very least, he knows which brand of vitamin he will buy when he goes to the health food store.

Even without a large group of people participating, the format of McCauley's study does not lack for statistical power. Studying individuals over time while varying their treatments is known as a crossover study. Studying patients' responses in sequence ideally allows all patients to act as their own controls.

But for McCauley, the most powerful achievement of the so-called citizen science effort was the fact that it could be done at all.

"We don't think we're going to prove anything from this except [that] we can do the experiment," he said. "It's not fancy science. It's real basic. But what's cool about it is, we have access to these things."

Before direct-to-consumer genetic testing, the idea of doing a DIY clinical trial based on variations in DNA would have seemed laughable. The ease of access to genetic information, owing to the low cost of gene sequencing, has quickly erased any memory of just how recently such data would have been a precious commodity. Few doubt that sequencing an entire human genome will become so inexpensive in the next few years that no technological barrier will exist to making a full-genome scan a part of every newborn's medical record. (Whether such scans will actually become routine in practice anytime soon seems a little doubtful. Witness the never-ending effort to make electronic medical records standard in doctors' offices and hospitals.)

Before whole-genome scans become universal, a smaller group of early adopters will likely seek out their own complete DNA data set, likely the same group who bought more limited scans from 23andMe, Navigenics, and others. Many will rely on paid services to interpret that information. Customers of 23andMe, after all, are not paying the company for an unfiltered list of As, Cs, Ts, and Gs. For $400, they get seamless online access through a slick Web interface to what those letters mean.

However they market themselves, personal genomics services do not sell the keys that unlock some deep secret. Their estimations of health risks and drug efficacy depend on published research accessible to anyone with an Internet connection. Compiling and making sense of those studies takes time and money. But anyone with the will and the dedication to apply themselves to understanding can find the same information if they want it. Already a Web site called SNPedia crowdsources much of the same information on the meaning of specific gene variations that gene-scanning companies charge customers to access. A free program called Promethease will interpret the results of your gene scan using SNPedia. For now, the most common sources of those scans are commercial services like 23andMe. But that could change quickly as the $1,000 genome looms.

As DNA sequencing gets even cheaper and open data sources like SNPedia become more robust, little will stop gene hackers like McCauley from playing with their genomes in whatever ways they wish. In that near future scenario, anyone could dig inside their genetic makeup for virtually no cost. McCauley and some friends have already developed a DIY genomics smartphone app that compares the SNPs analyzed by different gene-scanning companies and links to the research the companies use to interpret the results. Eventually, McCauley believes, he'll be able to upload his own genome directly into the app and have an open-source database tell him what the latest research says about his genes. He says this passion for understanding the genetic roots of his health does not make him a hypochondriac. "I don't think I'm so much the worried well. I'm a garage hacker, and I want to know how something works. But it's not the inside of a computer. It's me."

CHAPTER 12

Writing

At Mr. Gene, the basic building blocks of life are available at "unbelievable low prices." Shoppers can head over to mrgene.com and begin tapping out their custom sequences of As, Cs, Gs, and Ts as easily as updating their Facebook status.

To the uninitiated, the mere existence of DNA synthesis companies can feel like a dizzying trip into a science-fiction future that no one told you had already arrived. To take the basic stuff of our organic existence and add it to your online shopping cart for thirty-nine cents a letter seems brazenly dystopian. Who needs a metaphor to tell you life in the twenty-first century is cheap when the price tag is right there?

Of course, bored lab techs at any of the thousands of biotechnology companies around the world would roll their eyes at such high-toned rhetoric. Crunching DNA is what they do at work every day. Few pull into the office parking lot on a Monday morning in awe of their ability to play God. They just want to keep it interesting.

This is where Mr. Gene comes in. Synthesis companies allow biotech researchers to outsource the mundane, repetitive lab tasks that historically have consumed time and money at the expense of innovation. Someone has to fabricate the bits and pieces of DNA that every lab needs every time an experiment is tweaked. Services like Mr.

Gene allow scientists to point and click away their grunt work and stay focused on curing cancer.

The rise of cheap DNA synthesis has helped fuel the flip side of the genetic revolution spawned by sequencing. The technology to write DNA sequences has become cheap—not as cheap as reading DNA, but getting closer. As a result, creating a strand of DNA from scratch can be done simply by programming a computer, which instructs a small robot to mix the proper chemicals in the proper order. The ability to put DNA together letter by letter has led some scientists to embrace the possibility that one day soon they will be able to build new organisms never known to nature. Until now, genetic engineering has mostly meant snipping a gene or two from one microbe or mouse or frog and inserting that DNA into another cell, usually bacteria. Synthetic biologists want to break free from the idea that genes must originate in species. Instead, they want to piece them together based purely on a gene's function, regardless of where it originated in nature.

Parents who want to improve their kids' chances of growing up smart could do worse than name their offspring Chris Anderson. One Chris Anderson edits *Wired* magazine and popularized the idea of the "long tail"—the notion that businesses in the Internet era can just as easily target a large number of small markets as a small number of large markets. Another Chris Anderson curates the TED Conference, the exclusive California lecture series that captivates the global intelligentsia every spring. The third Chris Anderson may not be as well-known as the other two, but not because of any lack of comparable brain power.

This Chris Anderson goes by J. Chris Anderson. He is developing software that will power a robot that one day could build new forms of life.

Anderson's lab at the University of California, Berkeley, occupies a corner of the campus's grand new life sciences building, which was erected in hopes of attracting talent despite the budget woes that have shaken the country's most prestigious public university system. Having a lab in Stanley Hall means the university has high hopes for your work. Despite the financial strain, Berkeley remains a synthetic biology hub. Stanley Hall sits just down the hill from where Ernest Lawrence first smashed the atoms that would lead to the invention of the atomic bomb. Taking what at first seems like a modest discovery and using it to change history has happened at Berkeley before.

Anderson arrived at his office on a Tuesday afternoon dressed in Carhartt carpenter pants and plaid flannel. His desk was cluttered with scraps of paper and a stack of *Make* magazines, the bible of California DIY geekdom. As he talked, he fiddled with a chain made of dozens of small, identically shaped magnets. For more than two hours, as we talked, he unconsciously twisted and folded the rectangles into a mesmerizing assortment of shapes. The toy absorbed the leftover intellectual energy unable to find an adequate outlet in words as he sat behind his desk, clearly not the place where Anderson felt most at home. The hands of this thirty-three-year-old son of a former NASA Apollo engineer are clearly hands that need to play.

Anderson started his career in biotinkering as a protein engineer. Much of the discussion in biotechnology focuses on DNA, the instructional code that directs the building of all forms of life. But those instructions have an ultimate purpose: The construction of proteins that, joined together, become a living thing. Protein engineers insert themselves between DNA and the proteins they encode in an effort to create biological substances that never existed before. A successfully engineered protein "does something that is fundamentally unnatural," Anderson said. But Anderson was finding as the twentieth century ended and the twenty-first began that protein engineering was not living up to its promises.

Proteins are formed from chains of amino acids, twenty of which occur in nature. Each three-letter sequence of DNA codes for a specific amino acid; when cells translate these codes into the amino acids, protein synthesis occurs. The order and number of amino acids dictate the structure of the protein, thereby determining its function.

Proteins only become functional once they fold up into complex three-dimensional structures. Understanding protein folding is one of the most challenging problems in modern molecular biology. To engineer proteins requires a sophisticated understanding of that intricate folding machinery. The process is time consuming and fraught with potential for failure. Once that difficult work is done, the tweaked protein allows an engineer to force a cell to do one thing differently. For Anderson, changing just one thing no longer felt like enough. He was drawn to the just dawning possibilities of synthetic biology, which promised to allow bioengineers to make cells do several new things all at once.

Anderson's first big step into synthetic biology was an ambitious project. He wanted to engineer bacteria that a doctor could safely inject into the human bloodstream to target cancer cells and destroy them. In the process, he gained firsthand experience of just how far the analogy between synthetic biology and other kinds of engineering would take him. If synthetic biology was like mechanical engineering, Anderson was still trying to figure out how to machine the right nuts to go with the right bolts. If it were electrical engineering, Anderson was still trying to figure out how to solder together the wires, much less find the right wires to keep the lightbulb switched on.

"There are just so many tool limitations in this area that everything takes forever. It's very slow. It gets expensive mainly because it is so slow," Anderson told me. He called the process "agonizing."

Yet, when your aim is destroying tumors, you keep trying.

To get the bacteria to the tumors, they need to survive long enough in the bloodstream to reach their destination. He had read several studies suggesting that if he could hack a large set of genes into the bacteria's genome, he could increase the life span of his cancer bomb.

"It was just a lead. It wasn't like a solid 'this is going to work,'" Anderson recalled. But he and his team started building. The main challenge they faced was simply how to get the hacked bacteria to hold together—what he called "problems of assembly." Unlike other forms of engineering, in which you simply replace one part with another until you get the machine to work, each mistake in building a genetic machine means starting over from the beginning.

"It prevents you from just trying things, because there's so much of a commitment with just trying something," Anderson said. "It chews up time and money like nothing else to do assembly." In the end, Anderson's lab needed two years to successfully solve the many technical problems and build the bacteria with the genes they hoped would extend the microbe's half-life. Two years just to test one property, Anderson said ruefully.

"And then you find out two years later that it didn't even work."

Once the most basic problem of assembly is solved, the next one is regulation. In genetics, regulation refers to the ways in which cells determine whether a given gene within it is turned on or off and for how long. Researchers tend to think of cancer as a disease of regulation, in which the on-off switch shorts out and mutant cells grow out of control.

A more recent project in Anderson's lab has involved developing a cancer-fighting bacteria he calls a "payload delivery device." In Anderson's hypothesis, the genetically engineered bacteria invade the cancer cell, causing the cancer cell to surround the bacteria in a membrane called a vacuole. Another genetic tweak allows the bacteria to sense when it is in a vacuole, setting off a chain reaction that causes the bacteria and therefore the vacuole to explode. Ideally, the burst bacteria would release substances that would kill the cancer. The main challenge: regulating the amount of the particular protein needed to give the bacteria just the right "pop."

"You make too much of the popping enzymes and the thing is dead," Anderson said. "You make too few and it doesn't pop." His challenge became figuring out how to build in just the right balance.

In other, more established forms of engineering, there are often centuries of research behind the theory underlying the desired outcomes. Anderson likes to use a lightbulb analogy. Say you want the light to burn at a certain brightness. You have a battery, a bulb, and a resistor. A simple math equation will tell you what strength of each you will need to get the brightness you want.

In synthetic biology, that math does not exist. To figure out the arrangement of genetic parts needed to make the cancer-targeting bacteria burst in just the right way, Anderson had to go through the entire toolbox, starting over from scratch each time trying to find the proper parts.

Anderson's lab took two years to build five hundred different versions of the payload delivery device. "It's a lot of money," he says. But it worked.

The process was so painstaking, however, that Anderson realized a few more basic problems needed to be solved before going after big game like cancer. Hacking genes together had to be easier. So he decided to try to build a tool to do just that.

On a sunny, warm Memorial Day weekend, about two dozen people crammed into the Hacker Dojo garage in Mountain View to hear Anderson's pitch for coders who could design plug-ins for his synthetic biology software, called Clotho, after the youngest of the three Fates—the spinner of the thread of life.

Several of the faces clustered around the long folding tables were new to biohacking. Many were drawn by Eri Gentry, Tito Jankowski, and Joseph Jackson's pitch the week before at the Maker Faire. They represented a broad cross section of geekdom, from a middle-aged man getting a biology lab tech certificate at a community college to a computational biologist from the U.S. Department of Energy's Joint Genome Institute to a newly minted physics undergrad interested in biotech to a veteran Silicon Valley engineer looking for something new to fiddle with.

The fiddlers and tinkerers were exactly the crowd Anderson was trying to reach. He described how biotech lab work still involved doing too much by hand. Running a "manual shop" means work goes slowly. Slow means expensive. And expensive means projects can only become so complex before they become unaffordable. But manual does not have to remain the norm, he said. Computers and robots could likely automate every wet task in the genetic engineering handbook. "When it gets to that point, all people do is write software," Anderson told the group. He needed coders to get involved now, because most biologists trained in doing lab work by hand lacked the programming skills to set themselves free.

Anderson describes Clotho as "bioCAD" software, short for computer-aided design. Common uses of other CAD systems range from comparing colors of carpet in a three-dimensional simulation of your living room to designing microchips and spaceships. Anderson wants to use Clotho to design what biologists refer to modestly as "DNA constructs and strains," meaning unique strands of DNA not found in nature that do potentially world-altering things. He wants to build complex, genetically engineered machines, and he wants to make building them easy.

Still, the sheer complexity of genetic systems is preventing synthetic biology from becoming as straightforward as electrical engineering. Knowing what specific genes do and how they will react when combined with other genes requires tapping into galaxies of data that researchers have only started to chart. He doubts that building complex genetic machines via computer could be as straightforward as simply dragging and dropping parts around on a screen like designing a virtual IKEA kitchen. But Anderson says that even among life's most intricate systems, a little tinkering can take you somewhere.

"You're always running into things you don't entirely understand," Anderson told the meetup. But just like an electrical engineer going through a box of parts until he finds the right one to complete the circuit, tracking down the right gene sometimes just requires informed trial and error. "You can almost always, with even a little understanding, reduce things to where you can make at least one thing that does something. Intuition goes a long way."

Anderson doubts that amateurs without serious financial backing could build garage labs as sophisticated as those at a university. The main problem: keeping them clean. He recalled trying to grow orchids at home. He pointed out that he has plenty of experience setting up wet labs at universities designed to work with much more delicate organisms than even notoriously difficult flowers.

"I couldn't do it at home," he told me. "I could not actually create conditions that were sufficiently sterile. . . . As long as things are like that, it's going to be really hard to do anything in a garage."

But biohackers will not likely need their own labs within the next ten years, he said, unless they just want to do wet lab work for fun. Instead, Anderson predicts they will be able to do all the biohacking they desire in cafés, conjuring up genetic machines on their laptops using design software like Clotho.

"Where things start to change is when a lot of this stuff starts getting outsourced," Anderson said. Already customers can order up short strands of customized DNA from companies like Mr. Gene. Anderson says large-scale synthesis still costs too much for the average university lab or consumer, which is why his students still get stuck with the painstaking work of assembly.

Not only does he believe the cost will come down, but also that the second half of the process will also end up being outsourced. In any kind of genetic engineering, lab workers must follow up fabrication with analysis to make sure they have built what they hoped to build. So far, no companies offer cheap, standardized, centralized analysis to order. But Anderson believes start-ups or possibly some of the large biotech suppliers will start offering such services soon. As long as the price is right, he said, "there are no theoretical barriers to doing this sort of thing." If that happens, then the biohacker ideal of creativity and insight as the only barriers to entry in biotech could come closer to becoming a reality.

Anderson cites his brother-in-law, who invented a small electronic device that lets anyone turn a window-mounted air-conditioning unit into a cooling machine that transforms any insulated room into a walk-in refrigerator. The brother-in-law designed the device, called a CoolBot, but he did not build a machine shop to manufacture them in his garage. Instead, an entire infrastructure of electronics outsourcing services exists that allows his brother-in-law to get his CoolBots made for him—and make money selling them.

Anderson believes synthetic biology will become the same.

"To do it all by hand is really hard, and you need to do it for a long time before you're any good at it," he said. "I cannot possibly believe that this won't become a service." And unlike many scientists, he's not afraid to predict when he thinks comprehensive outsourcing of biotech wet lab work will become a reality. He thinks it will happen in fewer than ten years.

In the meantime, the scientists pushing synthetic biology forward with the most energy cannot relax at the bar after a tough day of gene hacking unless they have fake IDs. On a mild, sunny July afternoon, five UC Berkeley undergrads and their grad student adviser hunched over countertops in a windowless campus lab. Though summer break did not officially end for a month, the six would not be getting much sun. They had jobs that little more than a decade ago would have put them close to the bleeding edge of biotechnology. In a matter of weeks the team must hack together a genetic machine out of off-the-rack DNA. Their goal: to make a creature that no one has ever quite seen before in nature. Also, to make a creature cooler than anyone else's.

Since 2004, students like these have sacrificed fresh air and free time to take part in iGEM. The undergraduate contest that inspired biohackers like Mac Cowell has become a showcase for the gee-whiz advances in genetic engineering in recent years. In 2008, a team from Rice University developed bacteria to brew beer that contained resveratrol, a naturally occurring antioxidant in red wine that may or may not temper the effects of aging. That same year students from Slovenia took first prize for engineering a vaccine for the microbe that causes stomach ulcers.

More important than the individual inventions, however, is the underlying principle the contest's organizers at MIT hope to illustrate.

All contest entries must use genes taken from the Biobricks catalog standardized biological parts, each performing a specific known function, such as creating a fluorescent green protein, and each fits together in the same predictable way. BioBrick parts more than any other technological advance have served to push the idea that the same principles of engineering used to design mechanical and electronic devices can also be successfully applied to biology, making once radical feats of genetic manipulation so easy that even undergrads can create living machines.

Team leader Tim Hsiau, already a third-year graduate student in bioengineering at age twenty-three, explained Berkeley's 2010 iGEM entry to me and showed me the surprisingly simple equipment they were using to build their microbe. The team wanted to hack microorganisms known as choanoflagellates (koh-AH-no-FLA-juh-lates), planktonlike single-celled creatures that live in freshwater around the world and look like sperm cells. They are eukaryotes, which means their DNA is contained in a nucleus, as in human cells. (This sets them apart from most bacteria and other so-called prokaryotes, in which DNA floats freely in the cell.) The team told me that "choanos" come closer than any other organism to resembling animals without actually being animals themselves. Hsiau said that choanoflagellates have so far proven "genetically intractable," meaning that no one has been able to use the usual techniques of genetic engineering to alter their genomes. The Berkeley iGEM team wanted to be the first.

Hang around genetic engineers long enough and dazzling exercises in DNA hacking begin to sound routine. I found myself thinking that the Berkeley team's plan sounded promising in its "simplicity." If they couldn't tweak the choano's DNA directly, they proposed to stick the new genes in its food. Fortunately choanos feed on the most studied organism in the world. Genetic engineering may well not have existed if not for *E. coli* bacteria, the ubiquitous microbe used to hack DNA since the first genes were manually spliced. Hsiau said the team would splice into *E. coli* the genes intended for the choano and will have equipped the *E. coli* with the genetic equivalent of a detonator. When the choano eats the bacteria, a vacuole will envelope the prey in preparation for digestion. Triggered by a chemical cue, the *E. coli* bacteria will blow up themselves and the vacuoles, scattering protein and DNA inside the choano, like Anderson had done with his payload delivery bacteria. For their last trick, the team hoped to include as part of their DNA bomb a signal that would lead the choano to fold the nonnative genes into its own genome.

While interesting to scientists purely on principle, Hsiau said the ability to hack the choanoflagellate genome could have important uses in the future. Most practical uses of genetic engineering to create substances for humans still rely on *E. coli* to act as biofactories. But because *E. coli* is a prokaryote, it cannot make some proteins that could prove useful to humans, whose cells have nuclei. Because choanos are built more like human cells, Hsiau believes they also have the potential to make more substances that humans could find especially beneficial.

In genetic engineering, specialized proteins called restriction enzymes act as scissors to cut genes at specific places along the DNA strand. To isolate the specific sections of DNA they want to work with, biotechnicians rely on an arsenal of enzymes, each of which targets a different sequence of letters for cutting. To splice that segment of DNA into a different cell's DNA, the enzymes cut open the genome at a spot containing base pairs that match up with the segment being introduced. The foreign gene seals the gap, and a genetically modified organism is born.

To simplify the process of hacking multiple genes together, the BioBrick parts used by iGEM teams come precut in segments that all contain the same complementary sequences of letters at each end. In theory, eliminating the labor required to figure out how to prime each separate gene frees the genetic engineers to tinker with different designs the same way they would rearrange components on a circuit board.

Despite these successes, the sheer complexity of biology still stands in the way of creating the synthetic biology equivalent of Lego Mindstorms. Right now, the science is still largely in the Duplo stage.

"The problem that you have in looking at the type of part collection we're putting together is, at the end of that process you're inserting that system into a living cell. And that cell is more complex and less understood than we would like," said iGEM co-founder Tom Knight when I first talked to him in 2008 about mentoring future biohackers.

But Knight believes the strenuous effort necessary to improve that understanding will ultimately prove its worth. He told me he hopes to develop genetic machines that make exotic materials such as shatterproof ceramics and carbon nanotubes—graphite tubes the width of single molecules and up to fifty times as strong as steel. Companies, universities, and government researchers are racing to develop bacteria that can transform once valueless commodities such as algae and crop waste into biofuels.

Knight also predicted that engineered cells could make the manufacturing of physical materials as easy as moving information across the Internet. He compared the process to distributing text online.

"If you write an article, then it gets replicated and copied millions of times. The cost of these copies is very, very low. It takes very little in the way of infrastructure to do that. The bits are the bits. They flow across networks. They get replicated easily," he said.

Compare that to the "incredible" cost of making a car, he said. Especially consider the cost of making the factory that makes the car. Knight said he has an analogy he likes to share with his students: If automobiles were like living systems, the car you buy at the dealer would come with its own built-in factory that could make more. In living systems, the manufacturing technology and the objects themselves are the same.

"There's very little capital investment in building the next one. I think there's a good chance we can transition from a world where manufacturing is capital intensive to one where the cost of manufacturing is on the same par with the cost of replicating information."

In other words, he believes that, just as ones and zeros are to a computer, the As, Gs, Cs, and Ts of DNA can function as bits—the fundamental units of data processed by cellular machinery to create epic structures of logic such as ourselves. Bioengineers like Knight hope to be able to program cells like computers to interpret input into a desired output, except that unlike a computer, the cell could act as both the brain and the hands.

Emeryville, California, was once a rugged industrial town. Hemmed in by Oakland and Berkeley along the San Francisco Bay waterfront, the Sherwin-Williams paint factory once dominated the skyline here with its neon slogan WE COVER THE EARTH. The Judson Iron Works was one of the largest foundries in California. Those industries have long since left, along with the speakeasies, brothels, and gambling parlors that once led Chief Justice Earl Warren to call Emeryville "the rottenest city on the Pacific Coast."

Today a new kind of factory has taken up residence in town. Just up the street from Pixar Animation Studios and San Francisco's nearest IKEA, this assembly line occupies a few benches in a U.S. Department of Energy lab. Even when operating at full capacity, no one will be able to see what it churns out. No one will be able to buy anything, either. But if you get sick or drive a car, the plant managers here hope their little enterprise will one day change your life.

The BIOFAB was founded in late 2009 as the world's first open-source machine shop for synthetic biology. Founders Drew Endy and Adam Arkin believe the key to transforming biotechnology into an engineering discipline like any other lies in the creation of standardized parts. Like nuts and bolts, an automobile engine, or the components of an electronic circuit, these parts must fit together in predictable, measurable ways and do the same thing every time. As they get their factory up and running, Endy and Arkin will provide specs for these parts for free to universities and companies that want to build new things. They will also help design and create prototypes of new genetic machines made from BIOFAB parts.

But they have a problem. In other types of engineering, from furniture building to rocket science, math will tell you whether your parts will fit together and how they will work if they do. Build a table with legs that are too skinny and the table will collapse. Build a rocket with too small a fuel tank and the rocket falls to the ground without breaking free of earth's pull. Synthetic biologists do not have the math to predict how highly complex biological systems will behave when genetic parts are stitched together. For now they must put the pieces together first and then observe how they interact. This means that the BIOFAB must do more than build different-sized table legs. Factory workers must then fit each leg to the tabletop to see what happens. They must build the equivalent of thousands of rockets and send each up into the air.

This is the hard work of what biologists call characterization. As more is learned, Endy believes the BIOFAB will lead the way in closing the biggest gap in synthetic biology.

When brash biotech pioneer J. Craig Venter announced in May 2010 the creation of Synthia, the first self-replicating organism created entirely from DNA pieced together by humans letter by letter, the bacteria's genome ran about 1.1 million letters long. Venter's feat was a massive, expensive accomplishment in assembling DNA. But he did not take as huge a risk as the hoopla surrounding his accomplishment may have suggested, according to other synthetic biologists. He could believe with some confidence that when he "booted up" his creation, it would work, since Synthia was an exact replica of a bacteria that already exists in nature. But scientists have yet to build an entire microorganism letter by letter that has never been seen before in nature. The microbes manipulated by iGEMers contain genetic tweaks but remain mostly what nature made them. Most credible synthetic biologists will repeat again and again that the sheer complexity of even the most modest single-celled organisms makes the idea of a living thing engineered purely by human ingenuity a profound scientific challenge that has yet to be met. Venter's was an unprecedented feat of manufacturing. But evolution had done all the design work for him.

Endy likes to say that synthetic biologists today can readily engineer the genetic equivalent of a *New York Times* editorial—about a thousand characters. The longest functional genomes both built and designed by scientists run about twenty thousand matching pairs of letters—or "base pairs"—of DNA. At the far end of what's possible today, bioengineers are getting closer to the equivalent of a scientific paper, at more than thirty thousand characters. A reasonable short-term aspiration might be a one-act play, such as Jean-Paul Sartre's *No Exit*, at about one hundred thousand characters. To synthesize a microbe like *E. coli*, Endy says, the next generation of synthetic biologists will need the tools to write a Russian novel—each has about 3.5 million letters.

To help build a better typewriter, Endy calls standardization the first key step of the BIOFAB. To illustrate what he means, he looks to the ancient past. Endy told a group of bioengineers gathered at the BIOFAB that in Segovia, Spain, he visited a Roman aqueduct still standing strong after two thousand years. The multistory structure was made of what Endy called "standardized rocks": each one performing the same function in the overall design is the identical size and shape, allowing the entire edifice to hold together for two millennia with no mortar. The durability of the aqueduct stems from standardization, Endy said. This allowed those responsible for it to coordinate the construction work over time. One manager could pass down to the next the rules for making the rocks to replace any that failed. The manager need not leave actual rocks behind. Standardization also allows for the coordination of work among groups in different places, each an autonomous unit able to work together because each comes preprogrammed with the same rules for making their own rocks.

To create durable, useful, complex genetically engineered structures, the BIOFAB will need a lot of help from many others, Endy said. "If you can get people to work together over time and over different places, that lets you do things that are otherwise impossible."

When Endy thinks of the edge of the possible, he thinks of the village of Cherrapunji perched on the forested cliffs of the Khasi Hills in northeastern India. Known as the wettest place on the planet for its average yearly rainfall of 480 inches, Cherrapunji also attracts tourists with one of the world's more stunning examples of biomanufacturing. In a place teeming with water, bridges become a necessity. In Cherrapunji, the bridges are alive.

Along the banks of the region's many rivers, rubber fig trees grow shallow root systems that cling to the tops of rocks and other surfaces. Residents figured out that, using the trunks of other trees, they could guide these roots to grow straight and long—across to the opposite bank. Once on the other side, the roots grow into the ground, creating a sturdy span. As the roots continue to grow, the bridges become stronger. After the bridges are complete, a process that takes ten to fifteen years, their builders sometimes lay dirt and flagstones to ease the crossing.

Endy said the living bridges of Cherrapunji excite him because of the human capacity they demonstrate to engineer biology for practical, sustainable uses. (These bridges, he pointed out, also repair themselves and act to curb rather than exacerbate global warming.) He is also humbled by how little he says synthetic biologists can do with current technology to create something similarly marvelous. The bridges take about fifteen years to complete—a far shorter turnaround time than the new San Francisco–Oakland Bay Bridge under construction for decades a short distance from the BIOFAB.

Down a few floors in the same sleek office building, Amyris Inc. is a biotech company panning for gold in biofuels. Before that, founder

Jay Keasling had figured out a way to use the same genetic machinery researchers are working with to create fuels to make artemisinin, a well-known treatment for malaria. Artemisinin itself is nothing new; herbalists and pharmaceutical developers both have long extracted the compound from the Chinese sweet wormwood plant. But with its new synthetic cell, Amyris could brew the same amount of artemisinin in a few metal tanks as many farmers tilling the land. Using synthetic biology, the company could provide a reliably steady source of the drug not subject to the whims of nature that can afflict agriculture. They could control for quality—the cells made the exact same thing every time. And because these artemisinin factories replicate themselves, the supply is virtually guaranteed. Some enthusiasts have hailed Amyris as the best sign yet of synthetic biology's potential.

But the company's success has also made it a target. According to the company's critics, not everybody wins even when the prize is an unprecedented source of high-quality antimalarial drugs for some of the world's poorest people. That's because some of the world's poorest people also depend for a living on making and selling artemisinin. These farmers would seem inevitably outplayed in a marketplace where customers have access to a steady, quality-controlled supply of the compound brewed in vats in the United States or Europe rather than grown in remote African hinterlands. The convenience alone seems like a deal clincher. This has led to claims that Amyris, even though the artemisinin project is the nonprofit arm of its business, represents just another U.S. company with a technological silver bullet it will bestow in all its noblesse oblige on the rest of the world without asking people whether they really want the help.

Amyris chief science officer Jack Newman asserts that his company's artemisinin was never intended to supplant the supply grown in the ground. He points to reports that artemisinin farmers have suffered through several boom-and-bust cycles before Amyris was even a part of the marketplace. Newman sees his company's product as a way to stabilize the artemisinin economy so that farmers can count on a steady income rather than suffer the whiplash of shortage and glut. A stable supply also means a dependable source of medicine for those who need it, he says.

"During a shortage, for nineteen out of twenty kids that go to a clinic, there just won't be any medicine there," Newman told me. "And it will be 14 months that there won't be any medicine there."

Newman is a forty-three-year-old surfer who raises chickens in his Berkeley front yard. In his younger years he said he sat in redwoods as a member of Earth First! to keep loggers from cutting the trees down. He gave up on that kind of activism, but he says his passion for environmentalism still drives everything he does. Amyris's for-profit venture is biofuels. The company is currently processing Brazilian sugarcane using a genetically engineered microbe similar to the one used to brew artemisinin. Newman says that biofuels made using their process offer a 60 to 80 percent carbon footprint reduction compared to conventional fuels. And he does not apologize for the Amyris fuel's likely benefit for the company's bottom line.

"I think Amyris and myself personally are trying to find a way to effect change for the positive in the world while staying afloat as a business," he told me. "If you're cash flow negative and you're trying to do something good, you're going to go out of business. And if you're cash flow positive but aren't doing anything good, then what's the point?"

Despite Newman's assurances, it's not hard to understand how work being done by companies like Amyris can garner mistrust. Even setting aside for the moment the complex political economies of fuel production, biotech companies have a problem of aesthetics. Vats mounted in white rooms in blue-glass office buildings can become a kind of fetish object of the idea of progress. In such a setting, biology feels like nothing more or less than the next great industrial technology. It also feels like the culmination of a heroic intellectual effort, a tribute to the power of the rational. Yet such spaces also exist in a realm cut off from most of the rest of the world. This is obviously by necessity—bioreactors do not do as well around open windows. But a combination of factors has also largely limited the conversation around genetic engineering to the people doing it. This means issues and individuals like the artemisinin farmers can get left out.

One factor in closing off the discussion is the sheer sophistication of the science. While anyone who has used a computer can grasp the broad concepts behind synthetic biology, spend a few minutes around synthetic biologists talking shop or sharing diagrams and a layperson gets lost. Scientists sometimes have difficulty opening the conversation to outsiders because outsiders simply cannot understand them. That sophistication is both a product of and a contributor to the second factor, which is that nearly all biotech still takes place behind locked doors, in places where people need lanyards and photo ID badges or key codes and student credentials to enter. These barriers create the feeling that biotech belongs only to the select few: those who understand it and those who have access to the spaces where the work gets done, whether by virtue of knowledge or money. As a result, the issues that get discussed are those that matter most to biotechnologists. This is not to say that these scientists don't want to hear what others think. The other voices are simply harder to hear over the whirr of the centrifuge behind the locked door. Do-it-yourself biotech strives to bring molecular biology out of these closed-off spaces and give it to the public. Whether the public actually wants to have it is another question. Do-it-yourself biologists believe they should want it, if only because they have a right to it. Here, DIYers say. This is yours. Because DNA is us.

The dream of synthetic biology taps into deep wells of collective cultural longing that have been filled more typically by religion rather than science. Every mythology rests on a story of creation: where life came from, where we came from. But creation stories do not simply recount a sequence of events. They point to where, what, and whom we should worship. The human species throughout its existence has sought not only to understand and document the origin of creation but seems always to have instinctively revered the source of life, whatever the specific cultural guise. And whatever form that creative force takes, we have wanted to reunite with it. In most religious traditions, the pious heed the call to humble themselves as a way to channel that urge. In the great monotheistic religions, believers admit their powerlessness before God to receive His blessing. And nearly every form of worship involves some form of ritual sacrifice.

Yet stories and parables dating back at least to Icarus's fatal plunge speak to a different strategy. Some upstarts always try to get closer to the source of creation by ascending to the source's level. The story of Icarus is of course a parable about the folly of such an effort. Get too close to the sun and your hubris will get you burned. Yet in the eyes of twenty-first-century capitalist culture, which worships at the twin altars of the individual and technology, Icarus had initiative. And his melted wings do not represent some deep character flaw; he just needed better beta testers. Synthetic biology captures the imagination by promising to outdo Icarus, to get us closer to the source of life without ever leaving the ground.

Our power to mold the raw material of life has never been greater. Machines can stitch together custom strands of DNA letter by letter like a typewriter. Each advance makes the gene synthesis seem like less a problem of interest or technology but of access. As gene-writing becomes cheaper and more accurate, DIY biologists believe that designing an organism may take only as much technical know-how as using Photoshop or Microsoft Word.

As long as modern biotechnology has existed, descriptions of genetic manipulation have fallen back on the metaphor of playing God. The phrase was a cliché well before the first genes were spliced. The unleashing of atomic energy had long demolished any sense of religious proportion that had come before it. God's omnipotence lost a little sheen once humanity discovered it could engineer its own nuclear apocalypse—no divine intervention needed. The apocalyptic rhetoric returned, along with the idea that scientists were wrongly appropriating God's role after Boyer and Cohen invented recombinant DNA.

The playful, cheerful attitude favored by DIY biologists sometimes feels like an unwillingness to grapple with the primal unease biotechnology stirs. If DNA is just software and genetic machines just squishy versions of their silicon counterparts, then what's the big deal? Discomfort with the idea of tinkering with life seems like little more than a superstition to get over if DNA is just another raw material. But the feeling that life requires a unique kind of reverence hardly seems radical, nor is it necessarily religious. The history of technology on the one hand recounts the triumphs of human ingenuity over human suffering. At the same time, that history is littered with tales of environmental and human destruction: The erosion of biodiversity as a consequence of development and the insatiable appetite for natural resources. The poisoning of air and water by technologies that are supposed to make our lives better. Tinkering with life on an ecological level has often led to disaster. It is not irrational to worry that that tinkering on a genetic level could have similar consequences.

Yet the idea of building an organism has fascinated human beings since long before science existed to create science fiction out of, and that urge is not likely to be tamped down by the specter of microscopic Frankenstein's monsters. The iGEM contest skirts any sense of mythological awe at such an accomplishment in favor of the same geeky glee that might accompany building a model airplane. No sense of the sacred here.

The irreverence could stem in part from what ultimately amounts to an inferiority complex. No one knows better than genetic engineers how much their creations pale compared to the sublime intricacies engineered by nature itself. While undergraduates tinker with genes to make more nutritious beer, evolution has built the human brain. A bacteria tweaked to glow green may be cool, but as an engineering feat it is utterly primitive compared to the complex, robust, beautiful variations built by nature alone. Single-celled microorganisms have evolved since the dawn of life into foxes, Albert Einstein, and giant coastal redwoods. These undergraduate-forged DNA devices may use the materials that make life possible. But we are like toddlers pounding away on toy workbenches with wooden mallets compared to the master craftsmanship of nature's handiwork.

III

SAFETY/RISK

In May 2004, Steve Kurtz was driving to a Buffalo, New York, funeral home to recover his wife's body. Hope Kurtz was also his artistic collaborator. The two were preparing to unveil an installation at the Massachusetts Museum of Contemporary Art in North Adams called Free Range Grain. The piece involved setting up a simple molecular biology lab at the museum that would test everyday foods brought in by visitors to determine whether they were genetically modified. The subject was in keeping with Steve Kurtz's nearly two-decade-long artistic engagement with biotechnology as part of the Critical Art Ensemble, an internationally recognized group he had cofounded in 1987. Shortly before the exhibit was to open, Hope Kurtz collapsed from heart failure at the couple's Buffalo, New York, home. Kurtz called 911. Paramedics responded swiftly, but they could not save her. The police also responded to the call.

As Kurtz drove to the mortuary, he was pulled over by the FBI. From that moment, the next four years of his life would descend into a surreal nightmare. Instead of Kurtz retrieving his wife's body, federal agents would quarantine her corpse and test it for weaponized pathogens. The SUNY Buffalo art professor meanwhile was held for twenty-two hours without being read his rights. He was suspected of bioterrorism.

The Critical Art Ensemble has long focused on the complex issues raised by advances in biotechnology and genetic engineering. Their museum installations and performances typically use the lab tools of biotech as their media. Their goal has been to unsettle and provoke, but also to put the raw materials of biotech in front of the public to demystify a technology utterly hidden from view.

Buffalo police were troubled by what they saw at Kurtz's house, though they had no idea what they were really seeing. Petri dishes. Bacterial cultures. A mobile lab for testing whether food labeled organic truly was. In September 2002, five Americans of Yemeni descent were arrested just outside Buffalo and later convicted of providing support to al-Qaeda. The police did not know what was in Kurtz's house. But they did know bad things could happen in their city. They called the FBI.

The next day, while Kurtz was being detained, investigators from more than a half dozen federal, state, and local agencies raided his home. The group included the FBI, the Joint Terrorism Task Force, the Departments of Homeland Security and Defense, and Buffalo police and fire officers and New York state marshals. A half-block area around the house was cordoned off, and agents in haz-mat suits went inside. They seized his equipment, computers, manuscripts, books, and car. They also took his cat. Erie County public health officials condemned the house as a possible health hazard.

It took state health officials a week to test Kurtz's belongings before they announced that nothing in his home posed any kind of a health, environmental, or security risk. Only then could Kurtz retrieve his wife's body and return home.

But the findings did not end Kurtz's tangle with the U.S. Justice Department. Instead, his tribulations were only beginning. Federal law enforcement officials refused to drop his case. Initially, federal investigators sought to charge Kurtz under the USA Patriot Act's broad provisions for leveling accusations of bioterrorism. Under the 2001 law, anyone who "knowingly possesses any biological agent, toxin, or delivery system" that they do not intend to use for "a prophylactic, protective, bona fide research, or other peaceful purpose" can face up to ten years in prison. Kurtz had always argued that the purpose of his work is to critique and inspire discussion about the social, political, and philosophical implications of biotechnology. He argues that his bacteria are a form of protected speech. The implication of the indictment sought by the federal government is that Kurtz's work had no legitimate purpose. Legally, that defined his art as a threat.

A federal grand jury that was convened after Kurtz's detention rejected the government's call to label him a terrorist. But prosecutors were persuasive enough to prevent him from going back to his work unscathed. The grand jury instead handed down an indictment charging Kurtz with two counts of wire fraud and two counts of mail fraud—accusations typically leveled at organized crime figures. Also charged was Critical Art Ensemble collaborator Robert Ferrell. As the former head of the Department of Genetics at the University of Pittsburgh School of Public Health, Ferrell helped Kurtz obtain specimens for some of his pieces. The indictment claimed the two conspired under false pretenses to obtain bacterial samples from a supplier through that company's contract with the university. The government claimed that the bacteria were potentially harmful and that the pair had acted deceptively, because Kurtz as an individual could not have obtained the bacteria on his own. Prosecutors argued that the alleged crimes were still punishable under the Patriot Act. Kurtz faced a prison sentence of up to twenty years.

In 2008, a federal judge dismissed the government's indictment against him as "insufficient on its face," meaning that even if the charges against him were true, his actions did not rise to the level of a crime. Though artists around the world had rallied to Kurtz's defense, the case still took a severe toll, and not only on Kurtz himself. Although he was exonerated, his case still keeps biohackers on edge. Like him, they see the government's actions as motivated purely by politics. Anyone with even an undergraduate knowledge of microbiology could have ruled out Kurtz as a threat in short order, they say. Law enforcement's overreaction suggests to would-be DIY biologists that a petri dish makes anyone a suspect. How can anyone get any research done under that kind of pressure? And why would anyone take the risk?

CHAPTER 13

Threat

Aafia Siddiqui was born into an upper-middle-class Karachi family in 1972. Her father was a British-trained physician and her mother a social worker. Siddiqui came to the United States in 1991, where she remained for more than ten years. While in the United States, she earned an undergraduate degree in biology at MIT and a doctorate in neuroscience at Brandeis. While in school, she allegedly raised money for charities with ties to Islamist extremists. She returned to Pakistan in 2002. Shortly afterward, she became known as the most wanted woman in the world.

What happened to Siddiqui in Pakistan remains mired in secrecy and controversy. She was accused of having connections to al-Qaeda and now claims that U.S. authorities held her for years in a secret prison. The United States denies the allegation. Police arrested Siddiqui in Afghanistan's Ghazni province in 2008 outside the governor's compound. A federal grand jury indictment says that when she was picked up she was carrying handwritten notes referring to a "mass casualty attack" and listing several locations in New York City, including the Empire State Building, the Statue of Liberty, Wall Street, and the Brooklyn Bridge. The notes also allegedly described the construction of weapons of mass destruction, including bioweapons. The indictment claims Siddiqui was carrying "various chemicals" and a thumb drive containing correspondence that referred to attacks by specific "cells" and that put the United States on a list of "enemies."

In February 2010, a U.S. District Court jury in Manhattan convicted Siddiqui of attempted murder. According to federal prosecutors, a team of Army officers and FBI agents was preparing to interrogate Siddiqui after her 2008 arrest when she grabbed an officer's M4 rifle and opened fire. She did not hit anyone but was shot in the abdomen when the team fired back. During her trial Siddiqui took the stand against her attorneys' wishes and said she had never fired the gun, calling the charges against her "crazy." Though she was not accused of any terrorism-related offense, she also said that she did not know how to build weapons. "I don't know how to make a dirty bomb," she testified. "I couldn't kill a rat myself." The court sentenced her in September 2010 to eighty-six years in prison after her attorney pled for leniency, saying his client was mentally ill. During the sentencing hearing, Siddiqui disputed she was ill and said she had information that Israel was behind the attacks of September 11, 2001.

In a December 2008 report, the congressionally appointed bipartisan Commission on the Prevention of Weapons of Mass Destruction Proliferation and Terrorism pointed to Siddiqui as exactly the kind of scientifically literate, highly motivated extremist from whom the United States must do more to shield itself to prevent a potential bioterrorist attack. The report predicted that without urgent action, terrorists would use a weapon of mass destruction somewhere in the world by the end of 2013. And the weapon they would use would most likely be biological.

According to the report, the United States has invested most of its nonproliferation efforts since September 11, 2001, in preventing terrorists from obtaining nuclear materials. In the meantime, the report's authors wrote, "biotechnology has spread globally. At the same time that it has benefited humanity by enabling advances in medicine and in agriculture, it has also increased the availability of pathogens and technologies that can be used for sinister purposes."

In debates over bioterrorism and biotechnology, the key issue is the idea of "dual-use." Security analysts scrutinize biotechnology designed for a beneficial use to understand how readily it could be repurposed as a harmful weapon. The question spans much more than just the technical issue of what harm a genetically engineered microbe could do if released into the environment. Risk assessment also includes determining what skills and equipment a terrorist would need to make or modify a pathogen for malicious ends. With that knowledge, homeland security officials ideally could figure out who in the professional sphere already has the training and gear to inflict serious harm.

Still, despite even the post–September 11 anthrax attacks that federal authorities ultimately blamed on a U.S. Army biodefense researcher with access to heavily restricted pathogens, the monitoring of lab personnel still lags, according to the report. "The nuclear age began with a mushroom cloud—and all those who worked in the nuclear industry in any capacity, military or civilian, instantly understood that they must work and live under a clear and undeniable security mandate," the commission wrote. "But the life sciences community has never experienced a comparable iconic event to focus their attention on security."

If researchers in tightly controlled environments like U.S. military research labs require such high scrutiny, then what chance could a garage biologist have? Journalists covering the growing biopunk movement quickly embraced terrorism as its key threat. "The new danger next door?" was how the *San Francisco Chronicle*'s Web site teased its 2009 story on do-it-yourself biology.

"What's available to idealistic students, of course, would also be available to terrorists," Michael Specter wrote in a 2009 *New Yorker* story about the rise of synthetic biology. (Most of the story reflected Specter's enthusiasm for what he sees as synthetic biology's promising future.)

"The ability to create nasty pathogens like your hybrid rabies virus in your bathroom is becoming easier and easier," an unnamed federal official is quoted as saying in a 2009 *Homeland Security Today* story. "In the opinion of many in my field, this is much easier than trying to get enough fissile material to make a nuclear bomb and then being able to construct an effective bomb."

Since he first became interested in biohacking, DIYbio's Mac Cowell has felt like a self-conscious partner in a delicate dance with federal regulators and law enforcement. The Steve Kurtz case did not endear federal authorities to biohackers. The case was on Cowell's mind when he launched the DIYbio mailing list. In one of his first posts, he included a synopsis of Kurtz's ordeal as a cautionary tale to anyone in the United States contemplating wet lab work in their own homes.

This mistrust grew when early in 2009 about twenty people on the mailing list began receiving messages from members of an obscure consulting firm asking to interview them. They said they were consultants working on a workshop for regulators. The message's authors wanted to find out about trends in biohacking and where the movement was headed. "There was this sense they were going to open a dialogue with some government agency they couldn't name," Cowell said.

The dialogue never happened, according to Cowell. Instead, he believes the company was on a fishing expedition on behalf of federal homeland security officials. Some on the list suspected they were the FBI itself, trying to figure out just how dangerous these biohackers were. Could these seemingly harmless geeks who were extracting DNA in shot glasses and building gel boxes from spare parts be perfecting techniques that terrorists could use to build bioweapons? Were these biohackers terrorists themselves? Cowell believes the government's suspicions led authorities to miss an opportunity. (The FBI did not respond to interview requests.)

"I think that guys who do risk analysis for a living, they fundamentally are cynical," he said. "Any future they see is one that's already in the worst case. I felt like they dropped the ball a little bit by not starting the dialogue they promised."

But should the FBI even bother? Simpler, cheaper tools and streamlined techniques become available to everyone, not just those who will use them for good. The seemingly unwinnable war against e-mail spam or software viruses does not give confidence that somehow stopping their biotech equivalents will be any easier. If a garage scientist can engineer a new microbe to create a better biofuel or a cancer drug or a shatterproof ceramic, then why could a terrorist not create a stockpile of the worst germs ever to ravage humanity, adding in a few never before seen pathogens to spice up the mix? If we're already teetering on that precipice, what good will dialogue do?

In the riveting AMC television series *Breaking Bad*, actor Bryan Cranston plays Walter White, a high school chemistry teacher who finds out he has lung cancer and becomes a methamphetamine manufacturer to leave his family financially secure after he dies. Walt quickly excels at the trade, because his academic background allows him to produce a far more pure drug than the typical street cook can. Series creator Vince Gilligan scrupulously renders the details of Walt's lab setups—in an RV, locked in a basement—to show just how powerful one person can become with little more than the right know-how, a few everyday chemicals, a source of heat, and a few pieces of proper glassware.

Walt's success soon brings him into conflict with the hardened drug dealer Tuco in Albuquerque, where the show is set. Walt and his young partner, Jesse, decide they have no choice but to kill their rival, but neither has ever used a gun or had any real experience with violence. As the two argue over what to do, Walt pulls a small plastic baggie from his pocket. "I have a better idea," Walt says, holding out the bag to Jesse. "Beans."

"What are we going to do with them? Are we just going to grow a magic beanstalk, climb it, and escape?" Jesse asks, incredulous.

"We are going to process them into ricin," Walt says.

"Rice and beans?" Jesse says.

Though fictional, Walt and Jesse's plan brings some perspective to the debate over biohacking and terrorism. To hear biohackers tell it, a determined bioterrorist would hardly need a garage biotech lab to unleash mass casualties. Predictions of deadly genetically engineered germs have run parallel to every major development in biotechnology, from the first spliced genes in the 1970s to the latest synthetic biology advances. So far the predicted catastrophe has not happened (although opponents of genetically modified crops might call their spread a form of mass destruction). Vicious toxins like botulism and ricin occur in nature. They can be processed and concentrated with no more specialized tools than those found in the average kitchen. During its investigation of the post–9/11 anthrax attacks in the United States, the FBI said the perpetrator could have developed high-grade germs with $2,500 worth of equipment. In those fearsome cases, no genetic modifications are required.

In the same lab they use for cooking meth, Walt and Jesse process the castor beans into a fine powder, which they mix into a bag of meth they plan to offer to Tuco to snort. Walt tells Jesse about the most famous case of ricin poisoning, when in 1978 secret agents managed to use the tip of an umbrella to inject a minute ricin pellet into the leg of Bulgarian journalist Georgi Markov on a London street. Markov had spoken out against the Communist Bulgarian government. He died four days after the attack.

The idea that a few guys in a basement could cook up the same poison used by Cold War operatives may seem like typical television hyperbole. But the show's writers already had a real case to go on, a case where the culprit was far less believable than Walter White. In August 2008, fifty-seven-year-old Roger Von Bergendorff pleaded guilty to possessing a biological toxin. The unemployed graphic designer was staying in a Las Vegas motel near the Strip that February when he went to the hospital on Valentine's Day complaining of respiratory distress and soon fell into a coma.

When his cousin came to the hotel to clean out Bergendorff's things two weeks later, he found vials of ricin that authorities said contained enough poison to kill five hundred people. He also had guns in his room and an anarchist cookbook with the section on ricin production highlighted. Bergendorff's motive for having the poison has never been revealed. He had little to his name except $190,000 in debts reported in a bankruptcy filing. How could someone with Bergendorff's résumé end up at an Extended Stay America with enough ricin to kill an army battalion? It's really not that hard. Castor bean plants are commonly grown—and quite beautiful—ornamentals found in gardens across the United States.

In other words, genetic engineering is not necessary to commit an act of bioterror. Not only is it not necessary, it is also a much, much more difficult approach to bioterror than the simple processes available for manufacturing biotoxins. As Drew Endy does not hesitate to say: "We suck at engineering biology." Building a better bug is not the fastest way to world domination.

How far does synthetic biology have to go before building your own pathogen becomes as efficient and appealing as ricin? As of mid-2010, Endy says that 99 percent of designer DNA involves assembling strands fewer than twenty thousand letters long—far fewer than even the smallest microorganisms. (According to scientists with the Human Genome Project, the smallest free-living organisms are bacteria that contain about six hundred thousand letters in their genomes.) In testimony before a House Committee on Energy and Commerce hearing on synthetic biology in May 2010, Drew Endy made the case to committee chairman Henry Waxman and others that despite being labeled "extreme genetic engineering," the facts in the lab suggest that applying such a label to synthetic biology means more in theory than in practice.

Endy described his lab's work as an effort to understand how cells "make decisions, store information, and communicate." The lab's "holy grail," he said, was to create a DNA-based eight-bit information storage system—similar to a memory chip or a USB flash drive. But he said the genetic version would have two major differences. One was that the system would be made from proteins and DNA and function inside living cells. The other: "Our system will only store eight bits, which is eight billion times less than what you could store on an electronic memory stick available today from Walmart for twenty dollars." Endy explained that such a system could theoretically be used to count the number of times cells divide, and even act as a shut-off switch when cancer cells start replicating out of control. But Endy warned that such applications were a long way off.

"In total we need to design, build, and test about one hundred thousand base pairs, or letters, of highly engineered DNA," he said. "Using the best tools available it has taken us over one year to get the molecular pieces that comprise our first bit working."

The implications of Endy's remarks seem clear. Even at its most cutting edge, scientists cannot do anything with synthetic biology that could make it dangerous in the hands of terrorists (yet). Beyond the sheer challenge of piecing together a long enough genome, so many questions still exist about the basic biology of naturally occurring microorganisms that a living cell conjured purely by the human imagination remains out of reach of top researchers, much less extremists. That Pandora's box has not been opened.

Scientists less convinced by synthetic biology's potential have even stronger arguments against the need to fear some synthetic superbug. They point out that a synthetic organism's ability to survive in a petri dish says little about how it would fare in the real world. Evolution has spent eons forging the viruses and microbes that still make us sick today. They have survived because they are the fittest. Some skeptics of synthetic biology doubt the human intellect can somehow leapfrog the sheer ingenuity of natural selection when engineering a cell from scratch. Genetically modified organisms are a different story: Transgenic crops, for instance, have overtaken much of the United States's agricultural landscape. But genetically modified corn and cotton are still primarily products of nature, slightly tweaked. The idea that some mad scientist could with the tools available for the foreseeable future create a germ that could somehow outinfect the common cold or the seasonal flu strikes some scientists as the height of fantasy.

• • •

However unlikely the imminent danger of synthetic biology, biopunks still worry about how the public perceives them. Some resent the term "hacker" altogether and its association with malicious computer programmers. Others embrace their outsider status while fully believing they are setting themselves up to be placed on an FBI watch list. Nearly all share a libertarian ethic that holds outside regulation will stifle the creativity at the core of what they are striving to achieve. But they differ on how much of a threat their self-presentation poses to getting their work shut down.

That conflict boiled into public view when Chris Kelty first announced his Outlaw Biology? conference. An innocuous post to the DIYbio mailing list seeking suggestions for presentations quickly turned into an electronic slugfest. Some believed the name promised to attract sensationalist media coverage and undercover federal agents. Others said they felt no shame at celebrating their rogue status—that defiance of accepted norms was the essence of the biohacking ethic. Kelty said the name was meant to provoke discussion of what counts as legitimate biology.

A list member who identified himself only as Jake led the outcry against the name.

"This is sure to be a media disaster," Jake wrote.

"You might as well be holding up a sign asking the government to start tracking you. The FBI/CIA kept detailed records of *peace* protesters for christ sakes. You don't think they'll be keeping a better eye on 'outlaw biology'?"

In a later post, Jake raises the specter of a congressional crackdown:

"Believe me when I say that an anti-DIYbio bill is going to be a lot easier to pass when we self-relabel it the 'anti-outlaw-pirate-ninja-bio-hacker bill.' No politician in their right mind is going to raise any objection or even attempt to amend such a bill. And we'll have nobody to blame except ourselves."

Meredith Patterson was ready to take that risk. For her, the idea that DIY biologists would temper their public image to suit anyone violated the essential spirit of the movement. The laws are already stacked against any scientist who wants to work outside the entrenched institutions, she argues. Defiance is the default position of any do-it-yourselfer. In a post replying to the conference name controversy, she wrote: "You guys have known me long enough to know that I am a contrary bastard, so I'm only going to say this once: the more you call for self-censorship, the more flamboyant I'm going to get, because my dissenting voice is going to have to get louder and louder in order to be heard over your paranoia."

Some amazing feats of genetic engineering were on display at the iGEM jamboree in November 2009. But the most talked-about attendee may have been Supervisory Special Agent Edward You of the FBI's Weapons of Mass Destruction Directorate, Countermeasures Unit 1, Bioterrorism Prevention Program. Despite his overbearing title, You himself comes off not as stoic G-man but as a slightly more muscular version of the typical biogeek. Before joining the FBI, You worked as a gene therapy and cancer drug researcher for Amgen Inc., one of the world's largest biotech companies. He understands the science and has led the effort to convince biohackers not to think of the feds as members of the opposing team.

The FBI now has a weapons of mass destruction outreach program that among other things is working to build bridges from each field office's WMD agent to "canaries" in U.S. cities who could alert them to dangerous activities. The FBI clearly sees biohackers as essential members of the canary cadre's biosecurity division. "It was outreach, not oversight," You has said of his iGEM debut. "And it was blue jeans, not men in black."

You's effort to connect with the biohacker community began, in August 2009 in San Francisco where the FBI hosted its first-ever synthetic biology conference. At least one representative of DIYbio was asked to attend alongside industry and academic researchers. You followed up with a booth at the Outlaw Biology? conference. He genuinely seems to be trying to reach at least a modest understanding between himself and what computer security types would call the "white hats." Mackenzie Cowell and others in the self-identified DIYbio community have prioritized safety and openness in hopes of staying square with the government. They consciously set themselves up as seekers of beneficial knowledge. They are not the black hats, those in the software world who create computer viruses—and those in the biohacking world who would presumably create real ones.

At the same time, the white hats do not dispute the theoretical possibility of such danger. The tension arises over how much alarm such a possibility should raise versus the danger of cracking down on a fledgling technology that its proponents believe could deliver profound benefits to the world. According to You, the FBI itself has become sensitive to the need not to hinder innovation.

In its 2008 report, the federal antiterrorism commission paid much attention to the need to keep existing pathogens secure by securing the professional labs where they reside. The report also emphasizes the need for vigilance among scientists toward their own, especially those with access to microbes that pose a known threat. Noticeably absent: Any suggestion that the government should try to keep the tools of genetic engineering sequestered.

Perhaps this is why at the Outlaw Biology? conference, You and a few fellow agents saw no irony in setting up a recruiting station a few booths down from Meredith Patterson's demonstration of tabletop gene splicing. The most popular schwag at You's table was the magic marker–sized spray tube of hand sanitizer that Patterson used to create her ad hoc sterile environment.

You does not attend events like Outlaw Bio? and iGEM merely to reassure biohackers that they won't get into trouble. He is also working his beat. Like any city cop on the street, he can keep better tabs on the neighborhood the more the community knows and trusts him. You is the friendly face of an agency that, taken it at its word, wants to appear accessible in hopes of keeping the lines of communication open to a potential early warning system.

At Outlaw Biology?, You gave a presentation in which he emphasized that the federal government did not want to stifle research that could benefit society, even among researchers who were no longer keeping their work within the familiar boundaries of the university or the corporate lab. He assured biohackers that his office was a place they could turn for help if local authorities with less expertise began asking questions about whether their labs posed a threat. Though his name was not mentioned, the example of Steve Kurtz hovered over the proceedings, an example of the potentially severe consequences when those enforcing bioterrorism laws appear to lack scientific expertise.

You also appealed to biohackers' sense of civic responsibility. Much like the New York City subway system's "see something, say something" campaign asking riders to inform police of suspicious bags and passengers, You wanted do-it-yourself biologists to know that they were the eyes and ears in the best position to know if one of their own began veering the wrong way.

"What if there's an incident, even an accident, or even worse, an act of mischief or deliberate, intentional harm?" You asked later. "In the environment that we're dealing in now, you can imagine that there would be more legislation and regulation that could lead to increased restrictions."

You predicted such restrictions could be "knee-jerk" and "ill informed," leading to curbs on research, garage or otherwise.

"I can tell you right now that this is untenable even to the FBI. To the FBI this represents a national security risk as well," You said. "If you inhibit research, you're now stifling potential advances in the development of medicines, of vaccines, of countermeasures."

You's outreach efforts appear to be more than public relations. The Kurtz case alienated biohackers from the start. But a marginal group of nonconformist life-science enthusiasts hardly have the clout to force a major government agency to make amends for purely political reasons. You genuinely seems to see synthetic biologists, do-it-yourselfers, and other biotech iconoclasts as sentries on biosecurity's front lines.

"It really is up to the community, because you're the ones out there doing the work who know what the state of the art is," You said. Despite its in-house experts, You said that the FBI cannot hope to match DIYbio practitioners' "situational awareness." The same holds true for both law enforcement and policy makers, he said. "With the rate things are advancing, policy cannot keep up."

The subtext of You's message was clear: Law enforcement lacks the resources and even the mandate to fully monitor an underground movement not overtly engaged in anything criminal. The FBI also seems to accept that an idea like do-it-yourself biotech would not disappear even if law enforcement and legislators actively tried to kill it, an effort they have so far not seemed inclined to make with the exception of the Kurtz case. The ideas and technologies that inspire and enable biohacking—open source, Internet collaboration, standardized biological parts, DIY hardware, cheap DNA sequencing, outsourced DNA synthesis—are not themselves underground. These all have become mainstream elements of contemporary biotechnology and biomedical research. To outlaw biohacking would require heavily restricting all of these. In the end, the risk-benefit equation for which You serves as the government's front-line mathematician comes down to this: Does the risk of biotechnology's tools and techniques falling into the hands of terrorists outweigh the possible benefit of someone with access to the same tools and techniques curing cancer?

How law enforcement, security experts, legislators, and ultimately voters decide to answer that question could help or hinder the progress of biohacking as a serious movement. Whatever happens, diehards seem ready to fight for the right to pick up their pipettes. Said Eri Gentry: "If I imagine spending my time on anything else when I could be spending my time on saving people's lives, there is no comparison."

CHAPTER 14

Outbreak

In 1986, a small book created a big stir with its peculiar prediction of the apocalypse. In *Engines of Creation*, MIT engineering grad Eric Drexler described the coming nanotechnology revolution. Self-replicating machines the size of molecules would soon exist that could build almost anything by assembling individual atoms one by one, like the replicators in *Star Trek*. Machines operating on the molecular level could go into the body and heal damaged tissue. They could forge materials to build a new generation of spacecraft and mine precious resources on asteroids.

The field of nanotechnology has come a long way since Drexler's book, though so far not in the direction of his grandest predictions. Nanorobots are out; superstrong carbon nanotubes are in. But the book's most durable legacy has come in the form of the "gray goo" hypothesis. Drexler warned of a doomsday scenario in which self-replicating nanobots escape into the environment. Set loose in the world, the out-of-control machines outcompete weaker, less efficient living things and overrun the planet with untold copies of themselves—the dreaded gray goo.

Drexler described genetic engineering as the first concrete step toward the kind of molecular manipulation he envisions. More recently, in a talk with the futurist-leaning Edge Foundation, Drew

Endy described synthetic biology this way: "You have an actual living, reproducing machine; it's nanotechnology that works. It's not some Drexlarian fantasy." Because of these associations with nanotechnology as engineering on the molecular level, gray goo fears have haunted conversations about synthetic biology, as well.

Drexler himself later disavowed the danger of gray goo, saying that advances in the understanding of molecular manufacturing showed nanomachines did not need to have the ability to self-replicate. Synthetic organisms, on the other hand, are self-replicating by definition. Their power and importance as tools in industrial processes come from this ability to make more of themselves. This has led to worries about "green goo," which detractors of genetic engineering see as the more imminent and plausible possibility.

Jim Thomas is a leading voice among those who do not believe synthetic biology is worth the risk. What's more, he believes those risks are being undersold as the world's largest corporations look to synthetic biology as the next great tool to further their global dominance. A former Greenpeace activist, Thomas works for the Canada-based ETC Group, an international watchdog that warns of the environmental and human rights abuses enabled by biotechnology. An affable Brit with a fondness for sushi, slam poetry, and the history of science, Thomas may know more about the history and science of biotechnology than anyone else who thinks it's a bad idea.

After Craig Venter announced the creation of Synthia in May 2010, the first self-replicating cell made entirely by machine, the ETC Group called for a global moratorium on synthetic biology. ETC believes that until governments establish global rules for synthetic biology, all experimentation should halt.*

"We know that lab-created life-forms can escape and become biological weapons, and that their use threatens existing natural biodiver-

* The pioneers of recombinant DNA, Herb Boyer and Stanley Cohen, agreed to a similar moratorium on genetic engineering experimentation in the mid-1970s, though only after proving their invention worked.

sity," the group said at the time. Thomas worries about the possibility of medical or environmental catastrophe caused by a synthetic microbe escaping a poorly regulated wet lab. "Of course, those worries would be magnified if someone's doing that in their garage," he told me.

Thomas grounds his anxieties in basic ecology. Established ecosystems exist in a state of equilibrium. Each organism has its niche, its role to play as established by the long, subtle process of evolutionary adaptation. Invasive species such as zebra mussels arriving in the bilge water of cargo ships from Asia upset the ecological balance in a place like San Francisco Bay by crowding native species out of their niches. Whether because they lack a natural predator or overwhelm a native species's access to its food source, invasive species come to dominate ecosystems and are nearly impossible to eradicate.

In Thomas's worst-case scenario, synthetic microbes have the potential to become the ultimate invasive species. To understand why, he says, look no further than what researchers are already trying to design organisms to do.

The most talked-about biotech companies hoping to use synthetic biology to turn a profit are trying to engineer microbes to produce biofuels. As Thomas explains, the most effective microbes for that purpose would have the ability to break down cellulose, which forms the cell walls of green plants. As the most common organic material on the planet, cellulose for biofuels could come from nearly anywhere and require virtually no work to cultivate. A microbe that could process cellulose cheaply could turn grass into gold. Thomas imagines such an organism escaping into the environment with the power to devour any plant it encountered. What native species could compete with that?

Synthetic biologists also aspire to build organisms that can produce exotic materials. Perhaps a synthetic microbe has a unique set of genes that allow it to process some kind of cheap, naturally occurring material into a highly flexible or superrigid plastic. What happens when such a bug escapes from the lab and begins replicating itself in the soil of a forest or a farm? Does the ground underfoot turn to plastic, or at least become so permeated with the stuff that the land becomes a dead zone?

So far neither creature is believed to exist, but scientists who champion synthetic biology express no doubt that such inventions are possible. That enthusiasm alone is enough for Thomas: With such a possibility looming, the world needs to quickly formulate a way to deal with the potential danger. According to ETC's logic, scientists should stop their research to give the world time to figure out what to do.

At the time of the first hearing on synthetic biology held by the Presidential Commission for the Study of Bioethical Issues in July 2010, oil still gushed uncontrollably into the Gulf of Mexico from BP's deepwater well. Against that backdrop, Thomas's warning was stark: "Once these organisms are released, they cannot be taken back," he told the panel. "So this is a really big difference between a chemical spill or pollution, where it might be able to be cleaned up or degraded. It's just obvious fact that organisms that reproduce have the potential to be out in the environment forever. There's no way to find and kill every last one."

He reminded the panel of how rarely anyone manages to eradicate invasive animals or plants from an ecosystem once they take hold, much less a microbe. In the meantime, he says, even synthetic organisms designed to survive only in the lab could prove better able to adapt to the outside environment than anyone could predict. Perhaps they could even evolve and hybridize with other naturally occurring microbes, ensuring the engineered gene gets passed down and spreads to future generations of other organisms.

Some scientists accuse Thomas of fear mongering that owes more to science fiction than science fact. They compare synthetic cells to lab mice, which are bred and inbred for some specific research purpose and would never survive in the wild long enough to spread. Experimental organisms are created to survive only in one tightly controlled lab environment. The harsh unpredictability of nature would literally eat these pampered lab pets alive.

Yet Thomas's criticism extends beyond an elemental fear of pandemic or bioterror. His cause is more political than primal. For Thomas, the question of who controls biotechnology matters even more than any anxiety about biotechnology out of control. In his analysis, the history of technology demonstrates over and over again that those in power will always paint new technology as a panacea and then use it as a means to extend their power. As he puts it: "Powerful technology in an unjust world is likely to exacerbate the injustice."

Thomas prefers as a historical analogy the rise of what he calls synthetic chemistry. In 1828, the German chemist Frederick Wöhler combined two inorganic chemicals to create urea, a common organic compound found in urine. The discovery became known as the Wöhler synthesis. It launched the field of organic chemistry and struck a major blow against vitalism, the long-standing theory still prevalent at the time that claimed living matter contained some essential force or quality distinct from physics and chemistry. Within a few years Western European industrialists had established the modern chemical industry.

Thomas asked the president's bioethics commission to imagine if a similar panel were set up in 1828 and the same questions being asked about synthetic biology were asked then about chemistry: "Are synthetic chemists playing God? Will they make weapons of mass destruction? And will patents on synthetic chemistry lead to overbearing monopolies?"

History has answered those questions to what Thomas sees as his grim vindication. From the trenches of World War I to Auschwitz to napalm over Vietnam, chemical engineering has facilitated the conjuring of the most horrific arsenals ever known. The chemical industry also began with the work of tinkerers and enthusiasts, Thomas told me later. Now industrial chemistry is concentrated in the hands of a few massive multinationals. He reminded the panel of how long the spread of the chemical industry and its products continued before their impact on human health and the environment were taken seriously.

"That question didn't get an airing until 1962, with Rachel Carson's

Silent Spring," he told the panel. "And even when she did bring up these questions, she was vilified and attacked as an emotional and unscientific woman, as being an alarmist—just as cautionary voices on biotech are attacked today."

Thomas blames unchecked profit-seeking as the reason chemistry spiraled toward destructive purposes. He believes that without proper regulations, the same will happen with synthetic biology. He points to huge support for synthetic biology biofuel research from top oil companies, such as Exxon's $600 million deal with Synthetic Genomics, the Venter-run company that created Synthia. Under the partnership, Synthetic Genomics will work to create algae that can serve as a low-cost, high-yield alternative to petroleum. (Techniques already exist to process algae into biofuels, but Venter's company is working to engineer the algae to constantly secrete the oil straight out of its cell walls.) Thomas predicts that a biofuel economy on the scale of the present-day oil industry will transform the environment, the economy, and the political landscape to the same extent as petroleum a century earlier, and with what he sees as similarly negative consequences.

To their supporters in the United States, biofuels mean cleaner air, fewer greenhouse gas emissions and energy independence. To Thomas, they mean conflicts centered on three of life's essentials: land, water, and food. The coming bioeconomy will threaten access to all three for those who already struggle the most.

"Trying to guarantee the supply of sugar or cellulose or algae for the vats of synthetic organisms pumping out product will require a massive reorganization of natural resources, a grabbing of land and stripping away of plant matter and water and nutrients that could affect every part of the planet and some of the lives of the poorest people on the planet," he told the president's panel.

Ironically, Thomas in one respect does have strong common cause with DIY biologists, who support open access to the tools and techniques of genetic engineering as much as Thomas wants to restrict them. Both camps oppose corporate monopolies on biotechnology and the systems of intellectual property protection that support them.

For DIY biologists, gene patents and pay walls around research hinder access to knowledge that they could otherwise exploit to create innovative new technologies. For Thomas, gene patents mean corporate ownership of life itself and give companies the power to inflict biotechnology on impoverished people around the world nearly helpless to stop its encroachment. And he says DIYers who think that the unregulated practice of synthetic biology will mainly benefit garage innovators are being tremendously naïve about the consequences of unfettered corporate access to powerful technologies.

"We're going to get a lot of nasty surprises in the future," Thomas told me. "If synthetic biology is proliferating in amateur networks, we don't know where the surprises are going to come from."

Whenever I told nonscientists I was working on a book about people trying to figure out how to do genetic engineering in their garages, the first reaction was almost always a strained smile, followed by "Isn't that dangerous?" I would answer by laying out the arguments I'd heard pro and con: No, the science has not advanced far enough for garage hackers to create novel weaponized pathogens. Yes, some of the chemicals involved can make you sick and cause dangerous pollution if dumped down the drain. No, the people involved are working hard to develop safety protocols and understand that recklessness would lead to a quick government crackdown. Yes and no—bad guys with a semester of community college biology under their belts can get far more destruction for their dollar by whipping up a vat of botulism-causing bacteria in their basements than trying to splice genes. But the challenge for biohackers to date revolves much more around keeping their bugs alive at all than worrying about whether they'll escape and cause mayhem.

The more I explain, however, the more I sense that the explanation is beside the point. The question—"Isn't that dangerous?"—is not necessarily a request for more information or a rational assessment of the situation. Biology, particularly molecular biology, taps into our primal fears in a way that other sciences cannot. Physics deals in mostly invisible forces that feel somehow remote. Chemistry suggests bubbling potions that can burn or poison, but as long as we steer clear, we'll be okay. Biology, on the other hand, represents a threat with which we have all had intimate experience. Biology implies sickness, suffering, and death. Most of us have been hurt in some way by illness, as have our loved ones. The idea of tinkering with microorganisms—what we think of as the germs that hurt us, as we have been taught since childhood—seems at best to trivialize something we should approach with grave seriousness.

In a May 2010 story, *New York Times* reporters Andrew Pollack and Duff Wilson quote the then new director of the U.S. Occupational Safety and Health Administration, David Michaels: "Worker safety cannot be sacrificed on the altar of innovation. We have inadequate standards for workers exposed to infectious materials." In one case that received only a single sentence in the story, Pollack and Wilson tell of a researcher on leave from a U.S. biotech company working in New Zealand who becomes infected with the same meningococcal bacteria for which she was seeking a vaccine.

According to accounts in the New Zealand press, thirty-one-year-old British scientist Jeannette Adu-Bobie arrived in the country in 2005 to study meningococcal bacteria at a government-run Wellington research institute. In less than three weeks, she was stricken with blood poisoning caused by the bacteria. Doctors had to cut off both her legs and one of her arms to save her life. She also lost her fingers and thumb on her right hand, and her kidneys failed. The tragedy turned to scandal when the head of the research institute claimed investigators could find no fault in the lab or its protocols and that Adu-Bobie must have coincidentally contracted the rare disease somewhere else. Three years later the New Zealand government's labor department reversed its original conclusion and found that she likely was infected in the lab, though how she was infected remains unclear. "Dr. Adu-Bobie contracted meningitis, the same organism present in the laboratory," the final investigation stated. "There is no compelling evidence that this infection was contracted anywhere else."

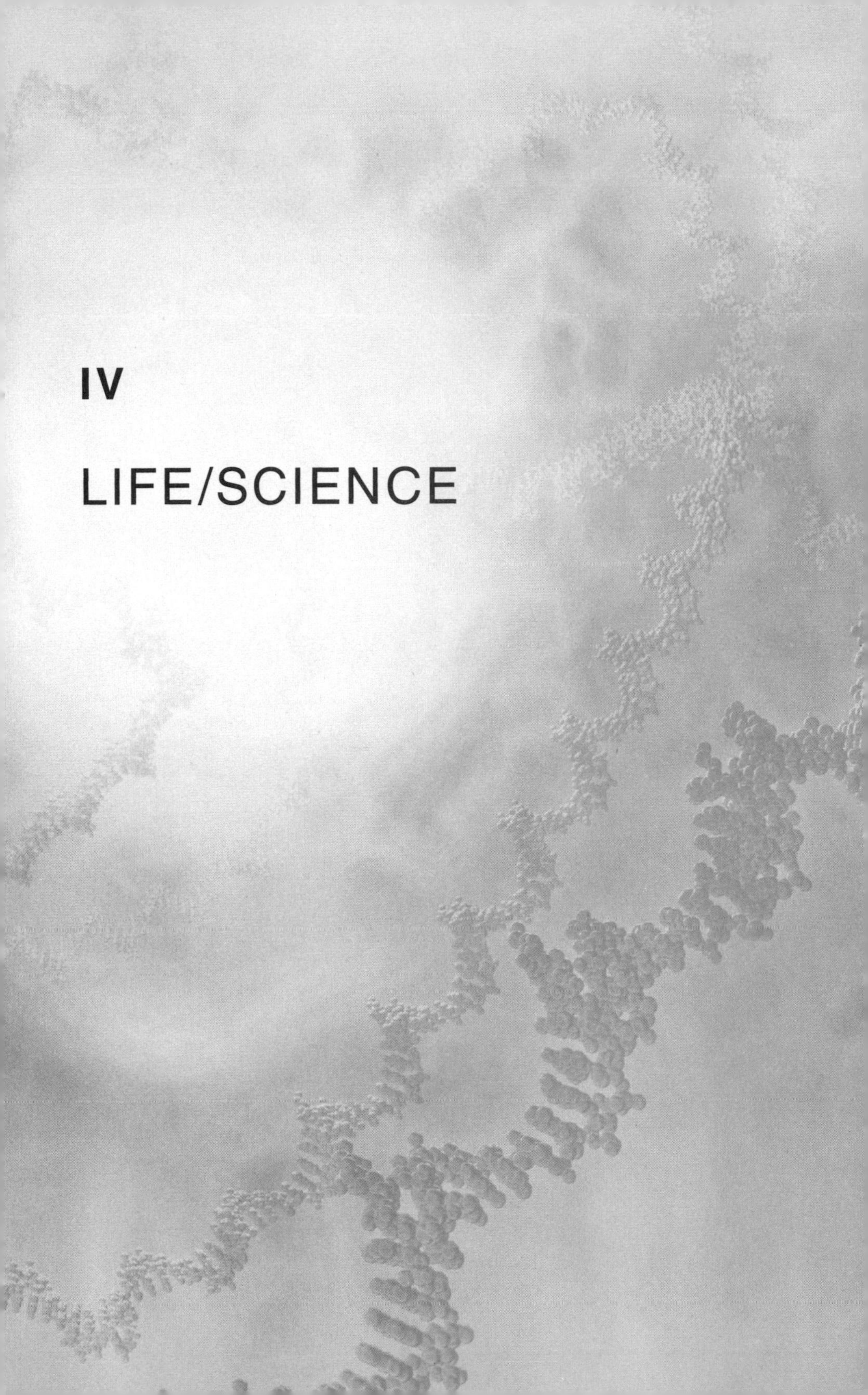

IV

LIFE/SCIENCE

If biohackers have a patron saint, his name is Freeman Dyson. A British math prodigy born in 1923, Dyson occupies a niche all his own along the insider-outsider scientist continuum. He became a member of the Royal Society before age thirty and spent more than forty years as a physics professor at Princeton. According to his own description, he has worked on nuclear reactors, solid state physics, ferromagnetism, astrophysics, and biology, "looking for problems where elegant mathematics could be usefully applied." Starting in the 1980s, he turned his eclectic scientific attentions to broader social issues in a series of popular books and lectures on topics ranging from war and peace to poverty, religion, and the origins of life. In recent years he has become well-known—and to some notorious—as the world's most impeccably credentialed global warming skeptic.

Nothing comes closer to a founding text of biohacking than "Our Biotech Future," an article Dyson wrote for *The New York Review of Books* in 2007. In it he predicts a world in which cheap, accessible biotechnology will lead to a world populated by fantastical living creations of our own design. Dyson contends that biotechnology will mean to the next fifty years what computers have meant to the last fifty. He predicts children will grow up playing biotech games the same way kids today play video games. A deep familiarity and comfort with what he calls Open Source biology will allow the next few generations to solve the world's problems using biotechnology in mind-bending ways. Superefficient crops that use less land and fewer resources because their leaves are made of silicon-based solar panels. Genetically engineered worms that extract metals from the earth, eliminating the need for conventional mining. Tree species that convert carbon dioxide and sunlight directly into liquid biofuels.

These biotech wonders will not only make life easier and the environment cleaner, but Dyson says they will also right an ancient inequality in the world's technological balance of power. He believes that a fundamental break in human history occurred about five thousand years ago, when the forging of bronze and iron began tilting hegemony away from societies dominated by agriculture. Economies based on biology became subservient to those based on chemistry and physics—what Dyson calls the gray technologies, such as wheeled vehicles, paved roads, and guns, which support the supremacy of cities. In Dyson's vision, the domestication of biotechnology as a grand source of new energy production will tip the balance back toward rural villages. Bioenergy will bring new wealth to the rural poor of countries like India, where he says impoverished villages will be transformed to resemble the gentrified villages of England, where subsistence farming has become a forgotten relic and highly skilled professionals connected to the global economy via the Internet become the norm.

"When industries and technologies are based on land and sunlight, they will bring employment and wealth to rural populations," Dyson writes. ". . . It is fortunate that sunlight is most abundant in tropical countries, where a large fraction of the world's people live and where rural poverty is most acute. Since sunlight is distributed more equitably than coal and oil, green technology can be a great equalizer, helping to narrow the gap between rich and poor countries."

Critics of Dyson's blithe vision abound. In a letter to *The New York Review of Books*, preservationist Wendell Berry called Dyson the latest in a line of soothsayers spanning back to the Industrial Revolution promising "the advent of yet another technological cure-all." Dyson fails to consider the social or environmental consequences of his remedies, a failure with all too many precedents.

Berry questions how domesticated biotechnology would serve the rural poor any better than any other form of industrialized agriculture, which Berry says concentrates wealth in the hands of corporations, not villagers. He also chastises Dyson for shirking the questions his own essay itself explicitly raises about the possibility that tweaking nature could have unforeseen and dangerous consequences. Berry lists history's "wish list of techno-scientific panaceas" that have also failed to pan out as promised: industrialization itself, eugenics, chemistry, nuclear power, the Green Revolution, television, the space program, and computers. "All those have been boosted, by prophets like Mr. Dyson, as benefits essentially without costs, assets without debits, in spite of their drawdown of necessary material and cultural resources. Such prophecies are in fact only sales talk—and sales talk, moreover, by sellers under no pressure to guarantee their products."

However realistic Dyson's predictions, his essay's influence has less to do with the specifics than the spirit of his predictions. At the core of the biohacker vision lives an extreme optimism about the power of biotechnology to do good. This optimism hinges on thinking about biology in terms borrowed from nonliving technology. Software. Hardware. Kits. Code. Chassis. Circuits. The closer bioengineers can come to making these metaphors literal in biotech, they imagine, the closer we all come to reaping the same benefits from biology that we do now from digital technology. In a sense, biohacker optimism about biotech has more to do with the undeniable transformations wrought by personal computers, mobile devices, and the Internet than any rigorous take on the current state of the life sciences. It's more optimism by analogy than by analysis: If computers have changed the world, and if we can make biology more like computers, then biotech will also change the world.

Developments over the past decade or so have invigorated believers in the idea that living organisms function very much like computers, and those who hope to manipulate life like others do silicon. The belief that cells function mainly as information-transmission devices underlies entire disciplines, such as bioengineering and synthetic biology. Computers have also gained ground in their ability to manipulate biology. Desktop DNA sequencers can read a genome and sync the results to an iPhone docked on the machine. Robotic arms manipulate minute amounts of liquid to splice genes and brew stem cells. Software lets designers piece together strands of DNA to build "genetic machines." The tools of information technology have become the tools of biotechnology. Under these circumstances, hacking digital technology and hacking biology start to feel like they might be the same thing.

Yet beyond Berry's lesson in the political history of technology, the science of biology itself could stymie anyone trying to think too literally about the similarities between self and software. Outspoken science blogger P. Z. Myers, a developmental neuroscientist at the University of Minnesota, believes the metaphor breaks down along a basic misunderstanding of the role of genes. "The genome is not the program; it's the data," Myers wrote in a 2010 riposte to the idea that superior computing power was going to allow researchers to reverse engineer the human brain within the next few decades.

"The program is the ontogeny of the organism," he continued, "which is an emergent property of interactions between the regulatory components of the genome and the environment, which uses that data to build species-specific properties of the organism." In other words, he believes that DNA by its very nature cannot function like computer code that performs specific predictable tasks in a specific predictable environment.

Under Myers's paradigm, a cell's means for determining what traits it will and won't express may yield varying interpretations of the same genetic code depending on where that cell lives. To adequately predict how a specific gene behaves in the same way one can dependably predict what will happen on a computer screen when the user presses enter would require accounting for every environment in which an organism might find itself. The layers of complexity generated by those indefinite environmental variables, compounded by what Myers shows to be the profound intricacies of the pathways that lead from DNA to self, give the impression that mapping those avenues is not just a problem of having enough computing power, but of understanding the basic qualitative difference in our scientific understanding of living and nonliving systems. In biology, we do not know everything we do not know.

Back at the Sprout lab in Somerville, the keg doesn't work. And the transhumanists are thirsty. Aubrey de Grey is there, famous beard down to his midsection, along with other luminaries of the radical life-extension movement. Yet after a day of presentations on enhancing intellect through implantable chips and uploading our brains to the Internet to achieve virtual immortality, all the assembled brainpower cannot figure out how to get the beer to flow.

The clunkiness of everyday technology (the keg) was an instructive counterpoint to the ultraoptimism of the 2010 H+ Summit. Since the 1980s, the transhumanist gospel has preached salvation through technology, of science as a way to achieve the ends that through much of history humanity reserved for religion: curing disease; the end of suffering; the fountain of youth; overcoming death. The movement believes not only in sustaining life, but also in using technology to enhance human bodies and minds. In the past, this has meant a strong faith in pharmaceuticals, nutritional supplements, exercise, and plastic surgery. This has also meant hope in cybernetics, artificial intelligence, and mind-computer interfaces.

The latest promises of genetic engineering have steered transhumanism toward DNA. The transhumanism on display in the summer of 2010 embraced an understanding of the self and of life as code. The same holds for bioengineering, though on a less grandiose scale. The crossover has brought some biopunks into the transhumanist fold, and vice versa, which to some might raise suspicions about biohacking as a whole. Like biopunks, transhumanists have always existed on the scientific fringe. Some are brilliant, some are cranks. Some are intellectually engaged, some are easily led. Some might dismiss biopunks as a whole by association: Of course they don't like institutions. When those institutions reject your ideas as bogus, it's always easier to blame the system than to reexamine the possible faults in your own thinking.

Yet most of the biopunks I met, whether transhumanists or not, had serious, practical criticisms of the way science is done. And their optimism about what genetic engineering can do in the near term is often shared by scientists in the academic mainstream. Is DNA like software? Are we like computers? The reasoning is clear: DNA encodes basic instructions—the machine language for all life. The letters are bits. Letters in sequence, regulated by a complex system of only partially understood RNA structures made from the same letters, encode for amino acids that join to form an endless variety of proteins. These in turn combine to form the tissues that make up an organism and enable the chemical processes that give the organism "life." Each layer builds upon the previous to create an ultimately logical structure. An engineer need only decipher that logic in order to reengineer it into something new and desirable.

Still, isn't there more to life than logic? Does living amount to no more than the exchange of energy across countless chemical processes? The debate over technological conceptions of human life has evolved in parallel with technology itself. As incremental technological advances pushed Western Europe toward the Industrial Revolution, incremental advances in life science brought scientists closer to an understanding of the cell as the basic organizational unit of all life. Cell theory undercut the idea that some supernatural or metaphysical force animated life in a way that was distinct from material existence, and therefore not observable according to any rational conception of science. The mechanistic view of the self began to emerge. As chemistry and physics came to dominate the sciences in the late nineteenth and twentieth centuries, the view of the self as a function of energetic forces, chemical interactions, and molecular processes came to dominate. In each case, as in our current moment, technology and culture created a feedback loop that skewed self-awareness toward a dominant metaphor.

Art has always taken the measure of that metaphor, now more self-consciously than ever. Contemporary art teems with deliberate engagements with the technological self and the unnatural other. Maybe because they already saw themselves as outsiders, and because what they make has no obligation to be useful, artists embraced DIY biotech for their own aesthetic ends well before science-minded biohackers got involved. Projects by Steve Kurtz and the Critical Art Ensemble are just a few of many that have brought petri dishes out of the lab and into the museum. Genes have become the paint and cells a new kind of canvas, with no snippet of DNA as eagerly embraced as the gene for green fluorescent protein.

No piece using the artificially amped-up version of a jellyfish gene that gives flouresence to living things attracted as much press as *GFP Bunny* by globe-trotting bioartist Eduardo Kac. In his 2000 "piece," French scientists bred an albino white rabbit with the gene for fluorescent green protein spliced into its genome. The rabbit still appeared white—unless it was illuminated with a very specific wavelength of light. Under such a lamp, the rabbit would glow green. A predictable outcry ensued about the exploitation of animals and the perversion of nature, which Kac tried to preempt by saying he considered all the debate surrounding the bunny part of the artwork itself.

The irony of much bioart lies in the creepiness or queasiness it induces: The more the "materials" used in the work resemble "us," the less comfortable we feel around it. Imagine a project in which living cells are draped on a frame and kept alive using fetal calf serum to grow into a jacket. An artist's collective known as the Tissue Culture & Art project created such a piece in 2004, titled *The Victimless Leather—a Prototype of a Stitch-less Jacket Grown in a Technoscientific 'Body.'* Now imagine the sleek lines of a modernist Bauhaus interior, a clean, minimally furnished space of glass and metal. Between the two, which provokes the gross-out reflex? These reactions speak to the same reflexive discomfort that underlies negative reactions to gene-splicing experiments of all kinds, but particularly to genetically modified foods. "Almost natural" upsets in a way that clearly unnatural does not. This is the uncanny valley biopunks are trying to leap as they strive to make more nonscientists comfortable with biotech.

Yet not all bioart falls into this breach. San Franicsco Bay Area bioartist Philip Ross spent a summer afternoon in a shaded Menlo Park backyard telling a group of DIY biologists about his own peculiarly elegant artistic manipulations of living things. Among his creations are a series of plug-and-play terrariums, self-contained systems for a single living plant that require no human intervention to survive except to plug the cord into the wall. He has also spent countless hours in his studio/lab perfecting what he calls "Mycotecture," in which fungi known as Reishi or Ling-chi mushrooms become the raw material for sculpture as they grow up within wooden frames to form tall, sturdy arches and walls. (He does this not through any kind of genetic engineering but through traditional mushroom-breeding techniques.) When the sculptures dry out, the mushrooms create surprisingly strong and durable bricks. At the end of an exhibition, patrons dismantle the sculptures by making the bricks into tea.

Ross said that talking about the aesthetics of biotech was tricky because the artworks themselves were alive. As such, they cross over the comfortable gap between subject and object, viewer and viewed, that characterizes conventional aesthetic experience. We have perhaps a little too much in common with what we're seeing to make a detached evaluation. "There isn't such a demarcation between the inside and the outside anymore," Ross said.

Andrew Hessel, the Pink Army Coop founder, comes across as more comfortable than most with the blurring of life and tech. He embraces the self as software. And he has a new girlfriend. "Her name is Cynthia," he purrs. "She's a little svelte."

Onscreen, she doesn't look like much. An artist's rendering shows a dark circle inside a lighter circle, while a ruler suggests her measurements at something less than one hundred micrometers, about the diameter of a human hair. Cynthia is in fact Synthia, the synthetic microbe assembled from scratch by J. Craig Venter in the spring of 2010. And Hessel is in love. But more like a grown-up suddenly smitten with a newborn.

"Suddenly all of you guys have the tools to become parents for bacterium," he tells his class of eager young breeders, a group brought together for their ripe intellects and fertile imaginations. The classroom is a carpeted 1970s-era meeting room with high ceilings and a weirdly angled asymmetrical shape. Just across the parking lot sits a defunct McDonald's with a skull-and-crossbones flag hanging in the front window, a decommissioned rocket just outside, and high-resolution tapes of various lunar landings within. Hessel is giving the keynote lecture wrapping up the biotech core curriculum at Singularity University, an unusual school being run in a small corner of the NASA Ames Research Center in Silicon Valley. The notion of the Singularity is the brainchild of the inventor and tech prognosticator Ray Kurzweil, who predicts that computers will begin to reproduce themselves around 2045, leading to an exponential explosion in artificial intelligence that will transform society in profound and profoundly unpredictable ways. (The term "singularity" was originally used to refer to the point in a black hole where the laws of physics break down.)

The concept of the Singularity has gained traction among some of Silicon Valley's biggest names, who have thrown their support behind the school. Among their many assertions, Singularity U.'s leaders believe the near future will bring many more Synthias of ever-greater complexity. Students are asked to prepare for these advances in "exponential technology" and be ready to take advantage of them for the greater good.

Hessel is dressed again in an untucked, black, button-down shirt over jeans. He gives the impression of a man too busy to sleep, a relaxed freneticism radiating from his red-framed glasses. A Canadian by birth and temperament, Hessel does not try to push this international crowd of young entrepreneurs to embrace his vision of a present in which humans now hold the reins of evolution. Instead, with his long Os and easy enthusiasm, he invites them to see what he believes are the wonder and wondrous opportunities represented by Synthia, the first self-replicating organism created entirely from DNA stitched together by computer.

"In my worldview, cells are computers. They compute. They literally process information. It's just not electronic information. It's not electrons. They process chemicals, chemical information," he said.

"I am a network of one hundred trillion cells working in close cooperation to make me," he said. "That's an amazing network. And that's why we have so much trouble figuring out how this network works. It's a four-billion-year development cycle. We're not going to figure it out in a few weeks."

The man-as-machine metaphor has existed for as long as machines have existed. Humans cannot help but see themselves in their creations. During the steam-power era, humans ran on fire, pistons, and cogs, a concept that suggested science could approach the body according to the same rational principles that applied to machines. Psychoanalysis arose as the use of electricity became widespread, when Freud saw mental health and conflict rooted in the flow of psychic energy. In the nuclear age, we became collections of atoms. As an assortment of particles, the body could be understood and healed most effectively through manipulation of those particles, through pharmaceuticals. In the digital age, some of us see ourselves as computers. And, as in every era, some scientists do not see the resemblance as metaphor. They view the similarity as evidence of sameness. As in every era, the recognition of this new identity—finally, we have glimpsed our true nature!—is seen by some as something to embrace. "Darwin did not see this coming. He had no idea that this would start to happen, that we would forward-engineer life," Hessel said. "Natural selection is over. It's done. It doesn't apply to us anymore."

Good-bye *Homo sapiens*, he said. Hello, *Homo evolutis*. This is a philosophical rather than scientific contention. Of course, natural selection is occurring every second of every day everywhere. Witness the rise of antibiotic-resistant bacterial infections as just the most unpleasant example. Even so, Hessel implies that science's ability to manipulate DNA to society's own ends takes some of the randomness out of evolution. Instead of a protracted process in which random mutations get filtered out over countless generations according to which proves the fittest, Hessel forsees biologists picking the mutations they prefer and trying to engineer fitness right into the first generation. If that becomes possible, biohacking will mean not only altering the DNA of individual organisms but hacking the evolutionary process itself. Of this profound potential change, Hessel says, "I think it will make us happy. I also think it will be a little weird."

Or a lot. Science fiction has dealt with these issues of technology and transcendence for much longer and with more nuance than most tech pundits. And science-fiction authors understand better than anyone else that visions of the future are always really about the present.

In the future conceived by British novelist M. John Harrison, most of the technology of which transhumanists dream has come to pass, but not with the results today's H+ aficionados anticipate. Blighted planets at the edge of the galaxy teem with urban dystopias, where the boundaries between life and technology have become impossibly blurred. Junkies lose themselves not to heroin but to tank farms, where they immerse themselves in neurochemical vats that feed them full-sensory fantasies for as long as the money lasts. Storefront "tailors" graft new bodies onto minds, and vice versa, the way someone today might get a boob job or a body piercing. A young girl's parents send her to become one with a spacecraft, her consciousness fused with the interstellar vessel while her broken body floats in a womb-like bath. Algorithms flit in the corners of bars, shadows untethered from the bodies that cast them. When spaceships crash, the code that ran their navigation systems comes oozing out of broken consoles and devours the human pilots, who now face a fate worse than death: They have become software.

Harrison's images are arresting, but he is not writing parables. The novels do not condemn technology, biological or otherwise, as the cause of social decay in his imagined worlds. His societies have the same moral contours as our own. Greed, vanity, despair, and the desire for transcendence drive his characters. His fantastical technologies merely allow these timeless flaws to inhabit new, grotesque forms.

Back on earth, a few miles down the road from Singularity U., a dozen cockroaches in a small plastic box milled around just like all cockroaches have milled around for tens of millions of years. These were not the standard-issue American cockroaches, the small brown nuisances with fluttery antennae that afflict every apartment in New York City. These were discoid cockroaches, also known as false death's head roaches for the skull-like markings on their thoraxes. As long as a grown man's hand is wide, these squat, brown pellets of armored ugliness leave you with little doubt that they could survive a nuclear blast if anything could.

Cockroaches have changed little from an evolutionary standpoint since their ancestors first appeared on the planet more than three hundred million years ago. As dinosaurs came and went and the first protohumans appeared, along with their increasingly complex centralized nervous systems, roaches remained creatures in which "thinking" occurs throughout the body. The slightest brush of air across the tiny bristles covering a cockroach's leg sends an electrical jolt to the abdominal ganglia, one of several control centers in the roach's body. A cockroach's escape response is hardwired into its abdominal ganglia. This neural circuit operates with extreme efficiency, allowing a roach to skitter out of the way as soon as it feels the rush of air from your oncoming shoe.

Tim Marzullo keeps his roaches from running off by sticking them in the freezer. When he brought them to the table in the plastic box, they looked sluggish, sliding around the bottom of the container on top of one another. He plucked one from the box and let it amble across the table while he explained the science of what he was about to do. The bug didn't get very far before he pinned it down. Eri Gentry wielded the scissors, and with a snip the cockroach was back in the bin one leg lighter. (Don't worry, Marzullo says. Legs grow back.)

Marzullo and Gentry then mounted the leg on a round pad atop a small metal box. From a wire jutting out of the box at one end, they pinned an electrode into the dismembered femur and switched the box on. Out of a speaker on the box next to the pad came a whoosh of static punctuated by clusters of pops that intensified when Marzullo brushed the bristles on the roach's leg with a small stick. This was, in a sense, the sound of a cockroach running away.

Even though the cockroach leg no longer has a body, the bug's nerve cells will stay primed for action until they dry out and die. So recently separated from its owner, the leg still has the electrochemical potential to send messages to the no longer attached ganglia that would send back the message to the leg to run away. The popping sounds overheard on what Marzullo has branded his SpikerBox are the electrical impulses being sent back and forth among the cells. Touching the bristles stimulates the cells, causing the electrical signal to spike.

Marzullo recently received his PhD in neuroscience from the University of Michigan. He said he knows firsthand the hoops aspiring neuroscientists typically need to jump through to get access to the expensive equipment typically used to measure action potential, which is what the SpikerBox does.

"Typically you have to bang on doors, apprentice, 'wash the dishes,' and learn how to use a $30,000 rig before you can finally do a neuroscience experiment when you're twenty-one on something you've been curious about since you were twelve," he says.

He and a colleague started Backyard Brains to give everyone the chance to tinker with neuroscience. They sell the assembled SpikerBoxes online for just under $100. They also sell a "bag of parts" kit inspired by the home-built radio kits of the 1960s and 1970s for $49.99. They hope to find customers among students, teachers, amateur scientists, biohackers, and other assorted nerds. But they also enjoy showing off the box to anyone who will spare them a minute. On the plane back from California, they pulled out the box (after informing the flight attendant and their seatmates) and performed what could be the world's first aerial cockroach brain recording ever.

Marzullo believes that even the most cutting-edge neuroscience still places the field at a time analogous to conventional medicine before anesthesia and penicillin. His grandfather died of extreme pain associated with neuralgia, a chronic nerve condition in which the perceived pain has no external cause. There is still much work left to do.

In these roaches, the brain is completely bound to the body. No one would mistake the crackles and pops of the cockroach's nerve endings for a soul, a common conflation humans make when we talk about our psyches. Marzullo's experiment grounds neuroscience in the physical rather than the mystical. The nervous system is something you can hear, something you can see, something that writhes around in a plastic bin and fills you with gleeful disgust. And something about which you want to know more. More minds engaged with neuroscience to Marzullo means more chances of solving the nervous system's mysteries and moving the science forward. To do that, Marzullo believes, scientists must persuade more people to abandon the notion that the brain somehow exists separate from the rest of the body in a way that puts it beyond scientific scrutiny. "What I want to do is change the perception that the brain is magical," he said.

Science can be slow. Science can be boring even as it yields some of the greatest wonders of human invention. Scientific decorum dictates a kind of reticence, a modesty that couches all public pronouncements in qualifiers. Emphasis is placed at least as much on what is not known as what is known. This slow, deliberative process ideally ensures that knowledge prevails over hype, that politics do not pollute science, that careerism does not compromise precision.

Yet that same process closes science off from the public. Never mind that the work happens behind closed lab doors. Basic science as practiced today is by necessity esoteric, and ever more so. Scientific knowledge advances in increments. Each one of those increments represents a greater specialization of knowledge. Professionals do all they can to keep up. The public has no chance.

Science journalists work to bridge this divide. Universities and companies have public relations departments. Hands-on museums like the California Academy of Sciences in San Francisco's Golden Gate Park offer breathtaking exhibits to keep the public engaged. But none of these is the thing itself.

Biopunks have not achieved any major scientific breakthroughs. Maybe they never will. But they all exhibit a goofy joy in what they do, like they're getting away with something. Because rather than wait for science to be done to them, they have decided to do science. No one else can tell them what they can and can't do. They will do it themselves. And with a little luck and talent, they might do something cool.

The Minutemen were one of southern California's iconic punk bands in the early 1980s. They formed in San Pedro, a blue-collar sprawl annexed by Los Angeles a century ago to ensure the larger city had a port. Guitarist D. Boon and bassist Mike Watt were underdog teenagers from an underdog town when they first heard punk. The pair quickly learned punk's lessons of fast, short, and loud. But they also picked up on an undercurrent that other bands missed. Punk was not just a sound. Punk was an ethic. Watt once described punk this way: "If it's something like, 'Everybody's telling me the wall's over there, but I'm going to push against it and see if it's really there'—to me, that's what punk is. An idealistic attitude."

Biopunks want to see if the wall around the fortress of Big Science is really as high as it seems—and whether the ticket price for entry through the well-protected gate is really as steep. A Minutemen motto was "We jam econo," a reference to the band's defiant commitment to making transformative music on the cheap. And they succeeded. Boon died at twenty-seven in a car accident. The band never sold a lot of records. But their songs have penetrated pop music's DNA.

As the Minutemen say: "Punk is whatever we make it to be."

Acknowledgments

This book never would have happened without the help, encouragement, forbearance, and support of my agent, Michael Bourret, my editor, Courtney Young, Maureen Cole, John Raess, Tim Reiterman, my friends and colleagues in the San Francisco bureau of The Associated Press, Michael Pollan, Emily Angell, Bob Calo, Will Harlan, my friends and colleagues from the University of California Berkely Graduate School of Journalism, Doug and Pat Bennett, Wayne Barrett, Jim Gorman, Alex Kazaks, Asa Muir-Harmony, Deborah Coffin, Gabe Westheimer, Kate M. Monroe, Ramona Bradley, Juno Turner, Miriam Casey, Andrew Wohlsen, Peter Wohlsen, Bob Wohlsen, Jr., the good people and coffee of Blue Bottle Coffee, Actual Café, Local 123, and Gaylord's Caffe Espresso, and Kim Bennett and Russell Wohlsen.

Notes

Epigraph

vii **"Every orchid or rose":** Freeman Dyson, "Our Biotech Future," *The New York Review of Books*, July 19, 2007, accessed September 22, 2010, http://www.nybooks.com/articles/archives/2007/jul/19/our-biotech-future.

vii **"The Commission believes":** Bob Graham and Jim Talent, et al., *World at Risk: The Report of the Commission on the Prevention of WMD Proliferation and Terrorism*, December 2008, accessed September 22, 2010, http://www.preventwmd.gov/world_at_risk_preface.

Preface

xi ***The Economist* ran a cover story:** "The RNA Revolution: Biology's Big Bang," *The Economist,* June 16, 2007.

I: Hack/Open

4 **Mackenzie Cowell issued this disgruntled tweet:** Mackenzie Cowell, @100ideas, April 26, 2009, https://twitter.com/100ideas/status/1625110043.

4 **took the Human Genome Project $2.7 billion:** "The Human Genome Project Completion: Frequently Asked Questions," National Human Genome Research Institute, National Institutes of Health, accessed September 22, 2010, http://www.genome.gov/11006943.

5 **letters of his DNA for $50,000:** Dmitry Pushkarev, Norma F. Neff, and Stephen R. Quake, "Single-molecule Sequencing of an Individual Human Genome," *Nature Biotechnology* 27 (2009): 847–50, published online August 10, 2009, doi:10.1038/nbt.1561.

6 **manufactured using millions of chicken eggs:** Karen Hopkin, "Egg Beaters: Flu Vaccine Makers Look Beyond the Chicken Egg," *Scientific American*, February 23, 2004, accessed September 22, 2010, http://www.scientificamerican.com/article.cfm?id=egg-beaters.

7 **a trivial effort to sequence the pathogen's entire genetic code:** Emily Singer, "Hunting for Clues in the Swine Flu Genetic Code," *Technology*

Review, April 29, 2009, accessed September 22, 2010, http://www.technologyreview.com/biomedicine/22569/?a=f.

Chapter 1: Blood/Simple

10 **the first to submit an order for smallpox:** Emily Singer, "Keeping Synthetic Biology Away from Terrorists," *Technology Review,* July 6, 2006, accessed September 22, 2010, http://www.technologyreview.com/biomedicine/17122/?a=f.

10 **could have spent her night shifts making polio:** Andrew Pollack, "Scientists Create a Live Polio Virus," *New York Times,* July 12, 2002, accessed September 22, 2010, http://www.nytimes.com/2002/07/12/science/12POLI.html.

10 **that army biodefense researcher Bruce Ivins:** "Justice Department and FBI Announce Formal Conclusion of Investigation into 2001 Anthrax Attacks," U.S. Department of Justice, accessed September 22, 2010, http://www.justice.gov/opa/pr/2010/February/10-nsd-166.html.

Chapter 2: Outsider Innovation

19 **the Registry of Standard Biologial Parts:** http://partsregistry.org/Main_Page.

20 **tried to build a genetic chemical sensor:** "Davidson College Synth-Aces," OpenWetWare, http://openwetware.org/wiki/Davidson:Davidson_2005.

20 **an e-mail list with fourteen hundred members:** To read or join, go to http://groups.google.com/group/diybio/.

22 **call what they do community-driven science:** "What We Do," Sprout & Co., http://thesprouts.org/what.

23 **acclaimed chemist George Whitesides:** For an in-depth explanation of microfluidics, see the Whitesides Research Group's Web site, http://gmwgroup.harvard.edu/research_microfluidics.html.

24 **described . . . the problem of explaining to his father:** Entry posted to Life Blog, December 1, 2009, http://www.molecularist.com/lifeblog/2009/12/increasing-tinkerability-explaining-diybio.html.

24 **"simplifying and domesticating":** Life Blog, December 1, 2009.

27 **"When we happen upon a technology":** W. Brian Arthur, *The Nature of Technology: What It Is and How It Evolves* (New York: Free Press, 2009), 11.

Chapter 3: Amateurish

29 ***Micrographia: Or Some Physiological Descriptions:*** For a beautiful online rendering of the *Micrographia,* visit the National Institutes of Health's Web site, http://archive.nlm.nih.gov/proj/ttp/flash/hooke/hooke.html.

30 **Mendel was born in 1822, the son of a farmer:** See Robin Marantz Henig's *The Monk in the Garden: The Lost and Found Genius of Gregor Mendel, the Father of Genetics* (New York: Mariner Books, 2001) for a richly detailed account of Mendel's early life.

30 **a full morning's journey away by train:** Henig, *The Monk in the Garden,* 48.

30 **test-taking anxiety:** For instance, Michael R. Cummings, *Human Heredity: Principles and Issues* (Pacific Grove, CA: Brooks Cole, 2008), 46.

30 lacked a real university education: Henig, *The Monk in the Garden*, 51.

30 Scholars debate whether Mendel fully grasped: See Allan Franklin, et al., *Ending the Mendel-Fisher Controversy* (Pittsburgh: University of Pittsburgh Press, 2008) for an in-depth analysis of the debate over challenges to Mendel's findings.

31 "uncertified substitute teacher": Henig, *The Monk in the Garden*, 61.

32 one of only five nonroyal British subjects: Stuart Jeffries, "The Question: Who Gets a State Funeral?" *The Guardian*, July 15, 2008, accessed September 22, 2010, http://www.guardian.co.uk/politics/2008/jul/15/past.margaretthatcher.

32 Morgan and his "fly-boys" were hackers: Christopher Kelty, "Meanings of Participation: Outlaw Biology?" accessed September 22, 2010, http://outlawbiology.net/about/wtf/.

33 a scientific prodigy who had earned his PhD in zoology by age twenty-two: From the biography of James Watson, the official Web site of the Nobel Prize, http://nobelprize.org/nobel_prizes/medicine/laureates/1962/watson-bio.html.

34 he had not finished his thesis by his thirties: Michel Morange, *A History of Molecular Biology* (Cambridge: Harvard University Press, 2000), 105–6.

34 passed the young, brilliant scientist's unpublished images: David Ardell, "Rosalind Franklin: (1920–58)," Web site of the National Health Museum, Atlanta, Georgia, accessed September 22, 2010, http://www.accessexcellence.org/MTC/.

Chapter 4: Make/Do

37 "the language of God": Francis Collins, *The Language of God: A Scientist Presents Evidence for Belief* (New York: Free Press, 2006).

38 Chinese dairy producers had been cutting their milk: Gong Jing and Liu Jingjing, "Spilling the Blame for China's Milk Crisis," *Caijing*, October 10, 2008, accessed September 22, 2010, http://english.caijing.com.cn/2008-10-10/110019183.html.

38 A Chinese court ultimately sentenced: "China Executes Two Over tainted Milk Powder Scandal," BBC, November 24, 2009, accessed September 22, 2010, http://news.bbc.co.uk/2/hi/asia-pacific/8375638.stm.

39 The FDA's own testing would reveal: Martha Mendoza, "FDA Finds Melamine, Byproduct in More Formula," The Associated Press, January 7, 2009, accessed September 22, 2010, http://www.seattlepi.com/national/395180_melamine08.html.

41 Over time the tinkerer came to be portrayed: For an in-depth analysis of the figure of the tinkerer in literature, see Mary Burke, *'Tinkers': Synge and the Cultural History of the Irish Traveller* (Oxford: Oxford University Press, 2009).

42 François Jacob famously personified evolution . . . Uri Alon elaborated on Jacob's ideas: Uri Alon, "Biological Networks: The Tinkerer as an Engineer," *Science* 301, September 26, 2003, 1866–67.

42 "Rather than planning structures in advance": Ibid.

44 Patterson debuted "A Biopunk Manifesto": The manifesto is available in its entirety on Patterson's blog, http://maradydd.livejournal.com/496085.html.

45 **summed up in 1993 by activist Eric Hughes in "A Cypherpunk's Manifesto":** "A Cypherpunk's Manifesto" is available online at http://www.activism.net/cypherpunk/manifesto.html.

46 **Johns Hopkins–trained geneticist and physician Hugh Rienhoff:** For more on Rienhoff's project, see Brendan I. Koerner, "DIY DNA: One Father's Attempt to Hack his Daughter's Genetic Code," *Wired*, January 19, 2009, accessed on September 22, 2010, http://www.wired.com/medtech/genetics/magazine/17-02/ff_diygenetics.

46 **about 200 million letters compared to 3 billion:** See "Extract DNA from Strawberries," a March 20, 2009, blog post on DIYbio.org for a video on how to extract DNA from strawberries, http://diybio.org/2009/03/20/extract-dna-from-strawberries/.

Chapter 5: Field Testing

49 **In 1989, doctors began encountering:** "Venezuelan Hemorrhagic Fever (VHF)," *Pan American Health Organization: Epidemiological Bulletin* 16:3, September 1995, accessed September 22, 2010, http://www.paho.org/english/sha/epibul_95-98/be953vhf.htm.

49 **tens of thousands of cases:** Pan American Health Organization data, accessed September 22, 2010, http://www.paho.org/english/ad/dpc/cd/dengue.htm.

49 **Scientists working in the Venezuelan hinterlands:** "Venezuelan Hemorrhagic Fever (VHF)," September 1995.

50 **has seen clashes between pro- and anti-Chavez students turn violent:** Zachary Lown, "Violent Student Protests Shut Down Parts of Mérida, Venezuela," venezuelanalysis.com, May 1, 2009, accessed September 22, 2010, http://venezuelanalysis.com/news/4411.

51 **Chagas disease is endemic:** "Chagas Disease (American trypanosomiasis)," World Health Organization fact sheet, accessed September 22, 2010, http://www.who.int/mediacentre/factsheets/fs340/en/.

51 **a standard technique known as an ELISA:** For more information on ELISA, see "Introduction to ELISA Activity," an animated tutorial from the University of Arizona, http://www.biology.arizona.edu/immunology/activities/elisa/main.html.

53 **Benkler wrote that the greatest drag on progress:** Yochai Benkler, *The Wealth of Networks: How Social Production Transforms Markets and Freedom* (New Haven: Yale University Press, 2006), available as a free PDF download, http://www.benkler.org/Benkler_Wealth_Of_Networks.pdf.

53 **"commons-based peer production":** Ibid., 72.

53 **IBM in 2003 made $2 billion:** Ibid., 47.

54 **The LavaAmp prototype is made from sheet metal:** For photos and video of the LavaAmp, go to http://www.lava-amp.com/technology.html.

55 **"We were allowed free, unsupervised access to the chemistry lab":** Kary Mullis, Nobel Lecture, December 8, 1993, http://nobelprize.org/nobel_prizes/chemistry/laureates/1993/mullis-lecture.html.

57 **Genentech's top-selling Avastin anticancer drug can cost nearly $100,000:** Gina Kolata and Andrew Pollack, "Costly Cancer Drug Offers Hope, but Also a Dilemma," *New York Times*, July 6, 2008, accessed on September 22, 2010, http://www.nytimes.com/2008/07/06/health/06avastin.html.

57 charging $200,000 a year for drugs that treat extremely rare diseases: Andrew Pollack, "Genzyme Drug Shortage Leaves Users Feeling Betrayed," *New York Times*, April 15, 2010, accessed September 22, 2010, http://www.nytimes.com/2010/04/16/business/16genzyme.html.

Chapter 6: Cheap Is Life

62 Pearl Biotech makes easy-to-track gel boxes: http://www.pearlbiotech.com/.

63 Internet Explorer's market share has eroded: Browser market share data via NetMarketshare, accessed September 23, 2010, http://www.netmarketshare.com/browser-market-share.aspx.

63 have reported that their API traffic is double their Web traffic: Nicholas Carlson, "What It's Like Working for Twitter," *Business Insider*, May 22, 2009, accessed September 23, 2010, http://www.businessinsider.com/what-its-like-working-for-twitter-clip-2009-5.

64 Wang began fiddling with an Arduino: http://www.arduino.cc/.

65 performing PCR meant hours of moving test tubes back and forth: Peter Gwynne and Gary Heebner, "PCR and Cloning: A Technology for the 21st Century," *Science*, special advertising section, February 9, 2001, accessed September 23, 2010, http://www.sciencemag.org/products/pcr.dtl.

65 Enter OpenPCR: http://openpcr.org/.

66 "If you need $5,000, it's tough having $2,000": Kickstarter FAQ, http://www.kickstarter.com/help/faq.

67 Dale Dougherty . . . posted a link to OpenPCR's Kickstarter page on Twitter: Dale Dougherty, @dalepd, June 15, 2010, http://twitter.com/dalepd/status/16253152287.

67 A few hours later, tech publishing superstar Tim O'Reilly retweeted: Tim O'Reilly, @timoreilly, June 15, 2010, http://twitter.com/timoreilly/status/16268192528.

67 the pair had received more than $12,000: OpenPCR's Kickstarter page, accessed September 23, 2010, http://www.kickstarter.com/projects/930368578/openpcr-open-source-biotech-on-your-desktop.

68 More than half the acres of corn and cotton and more than 90 percent of the acres of soy: "Adoption of Genetically Engineered Crops in the U.S.," U.S. Department of Agriculture Economic Research Service data, accessed September 23, 2010, http://www.ers.usda.gov/Data/BiotechCrops/.

Chapter 7: Homegrown

70 This was one innovation that was clearly crowdsourced: Jared M. Diamond, *Guns, Germs, and Steel: The Fates of Human Societies* (New York: W. W. Norton, 1997), 105.

71 Some of the fiercest criticism came from India: See Vandana Shiva, et al., *Seeds of Suicide: The Ecological and Human Costs of Globalisation of Agriculture* (New Delhi: Research Foundation for Science, Technology and Ecology, 2002).

71 Sainath blamed India's embrace of the World Trade Organization: P. Sainath, "The Largest Wave of Suicides in History," *Counterpunch*, February 12, 2009, accessed on September 25, 2010, http://www.counterpunch.org/sainath02122009.html.

71 **Centrists faulted allegedly corrupt government bureaucracies:** Sanjeev Nayyar, "Killing with Kindness," *Business Standard*, February 2, 2007, accessed September 25, 2010 (registration required), http://www.business-standard.com/india/storypage.php?autono=273317.

72 **Monsanto has called those allegations "sensational" and "speculative":** "Indian Farmer Suicide—The Bottom Line," *Beyond the Rows* (Monsanto official blog), March 26, 2009, accessed September 25, 2010, http://www.monsantoblog.com/2009/03/26/indian-farmer-suicide-the-bottom-line/.

72 **the ruined fields "raised eyebrows":** Glenn Davis Stone, "The Birth and Death of Traditional Knowledge: Paradoxical Effects of Biotechnology in India," in *Biodiversity and the Law: Intellectual Property, Biotechnology and Traditional Knowledge*, ed. Charles McManis (London: Earthscan Publications, 2007), 226.

73 **But Navbharat never faced any legal blowback:** Ibid.

73 **Growers who are discovered:** Robert Kenner, director, *Food Inc.*, 2008.

73 **the biggest cotton-producing state in India:** The Cotton Corporation of India statistics, accessed September 25, 2010, http://www.cotcorp.gov.in/statistics.asp.

73 **"stealth seeds":** Ronald J. Herring, "Stealth Seeds: Bioproperty, Biosafety, Biopolitics," *Journal of Development Studies* 43:1, 130–57, January 2007.

74 **farmers in Gujarat have created a "cottage industry":** Ibid., 132.

75 **had to rely on the same dicey webs of reputation . . . BesT Cotton Seed:** Ibid., 134.

75 **"Stealth transgenics are [being] saved":** Ibid., 130.

75 **"Farmers pursue stealth seeds . . . often give better results":** Ibid., 146.

75 **can cost less than a third . . . outselling legal seeds by as much as ten to one:** Ibid., 135.

Chapter 8: My Life

77 **an unlikely alliance of bicoastal progressives and heartland religious conservatives:** John Schwartz and Andrew Pollack, "Judge Invalidates Human Gene Patent," *New York Times*, March 29, 2010, accessed September 25, 2010, http://www.nytimes.com/2010/03/30/business/30gene.html.

77 **U.S. District Court judge Robert Sweet's ruling against the U.S. Patent and Trademark Office:** *Association for Molecular Pathology, et al., v. United States Patent and Trademark Office, et al.* (N.D. Calif. 2010), accessed September 25, 2010, https://ecf.nysd.uscourts.gov/doc1/12717579719.

77 **about 20 percent of the human genome:** Schwartz and Pollack, "Judge Invalidates Human Gene Patent," March 29, 2010.

77 **more than $3,000 per patient:** Ibid.

78 **San Francisco (UCSF) microbiologist Herbert Boyer and Stanford researcher Stanley Cohen met at a Honolulu deli:** The University of California, Berkeley's Bancroft Library offers a nice thumbnail history of the founding of modern biotechnology. See "Biotechnology at 25 : The Founders," part of the Regional Oral History Office's Program in Bioscience and Biotechnology Studies exhibit, http://bancroft.berkeley.edu/Exhibits/Biotech/25.html.

78 **DOOMSDAY: TINKERING WITH LIFE:** "Doomsday: Tinkering with Life," *Time*, April 18, 1977, accessed September 25, 2010, http://www.time.com/time/magazine/article/0,9171,914901,00.html.

79 **Boyer harnessed his discovery to cofound Genentech:** "History," Genentech, accessed September 25, 2010, http://www.gene.com/gene/about/corporate/history/.

79 **In 1980, Genentech went public:** "Corporate Chronology," Genentech, accessed September 25, 2010, http://www.gene.com/gene/about/corporate/history/timeline.html.

79 **The Cohen-Boyer patent . . . became the gold standard:** For an extensive discussion of the implications of Cohen-Boyer, see Maryann P. Feldman, Alessandra Colaianni, and Connie Kang Liu, "Lessons from the Commercialization of the Cohen-Boyer Patents: The Stanford University Licensing Program," *Intellectual Property Management in Health and Agricultural Innovation: A Handbook of Best Practices*, eds. A. Krattiger, R. T. Mahoney, L. Nelsen, et al., (Davis, CA: PIPRA, 2007). U.S.A., accessed September 25, 2010, http://www.iphandbook.org/handbook/ch17/p22/.

79 **"Anything under the sun that is made by man":** *Diamond v. Chakrabarty*, 447 U.S. 303 (1980), accessed September 25, 2010, http://laws.findlaw.com/us/447/303.html.

79 **Bayh-Dole Act:** For an extensive history of the act with a positive spin, see "30 Bayh-Dole: Driving Innovation," http://www.b-d30.org/.

81 **more than $85 billion:** Alexander G. Higgins, "Drug deal: Roche buys Genentech for $46.8 billion," The Associated Press, March 12, 2009, accessed September 25, 2010, http://www.usatoday.com/money/industries/health/drugs/2009-03-12-roche-buys-genentech_N.htm.

81 **on annual revenues of more than $13 billion. . . . Its top seller, Avastin, brought the South San Francisco–based company nearly $2.7 billion:** "Genentech Announces Full Year And Fourth Quarter 2008 Results," Genentech, accessed September 25, 2010, http://www.gene.com/gene/news/press-releases/display.do?method=detail&id=11767&categoryid=2.

82 **one of these led to the development of Avastin:** "Napoleone Ferrara," Genentech, accessed September 25, 2010, http://www.gene.com/gene/research/sci-profiles/rsrchonc/tumbioangio/ferrara/avastin.html.

83 **the industry as a whole had finally turned a profit:** "Biotech Industry Turns a Profit for the First Time, But Milestone Overshadowed as Companies Struggle for Survival, Report Finds," Burrill and Company, February 25, 2009, accessed September 25, 2010, http://www.burrillandco.com/news-355-Biotech_Industry_Turns_a_Profit_for_the_First_Time_But_Milestone_Overshadowed_as_Companies_Struggle_for_Survival_Report_Finds.html.

85 **he founded the Pink Army Cooperative:** http://pinkarmy.org/.

Chapter 9: Ladies and Gentlemen

87 **"In beauty, or wit, / No mortal as yet / To question your empire has dar'd.":** Alexander Pope, "To Lady Mary Wortley Montagu," *The Poetical Works of Alexander Pope, Volume II* (London: William Pickering, 1831), accessed September 25, 2010, http://books.google.com/books

?id=tSM-AAAAYAAJ&lpg=PA189&ots=QR1V3NT2Ye&dq=pope%20%22to%20lady%20mary%20wortley%20montagu%22&pg=PP6#v=onepage&q=pope%20%22to%20lady%20mary%20wortley%20montagu%22&f=false.

88 **Lady Montagu contracted smallpox:** Stefan Riedel, M.D., PhD, "Edward Jenner and the History of Smallpox and Vaccination," *Baylor University Medical Center Proceedings* 18(1), January 2005, 21–25.

88 **as common to her era as cancer or heart disease:** "Edward Jenner: Smallpox and the Discovery of Vaccination," Doctor's Independent Network, accessed September 25, 2010, http://www.dinweb.org/dinweb/DIN Museum/Edward%20Jenner.asp.

88 **About 30 percent of those afflicted:** "Smallpox Disease Overview," Centers for Disease Control and Prevention, accessed September 25, 2010, http://www.bt.cdc.gov/agent/smallpox/overview/disease-facts.asp.

89 **a letter to a friend:** Lady Mary Wortley Montagu, *Letters of the Right Honourable Lady M—y W—y M—e: Written During her Travels in Europe, Asia and Africa . . . , vol. 1* (Aix: Anthony Henricy, 1796), 167–69; letter 36, to Mrs. S. C. from Adrianople, n.d., accessed September 25, 2010, http://www.fordham.edu/halsall/mod/montagu-smallpox.html.

90 **between 1 percent and 2 percent of patients deliberately infected:** "Variolation," U.S. National Library of Medicine, accessed September 25, 2010, http://www.nlm.nih.gov/exhibition/smallpox/sp_variolation.html.

90 **but found little interest among British physicians:** Meyer Friedman, M.D., and Gerald W. Friedland, M.D., *Medicine's Ten Greatest Discoveries* (New Haven: Yale University Press, 1998), 66.

90 **infamous rogue:** "Edward Wortley Montagu, 1713-1776," The Montague Millennium: 1000 Years of Worldwide Family History, accessed September 25, 2010, http://www.montaguemillennium.com/familyresearch/h_1776_edward.htm.

91 **inoculate her four-year-old daughter . . . and later exposure to smallpox among some proved they had become immune:** Riedel, "Edward Jenner and the History of Smallpox and Vaccination," 22.

91 **including the young Jenner:** "Edward Jenner: Smallpox and the Discovery of Vaccination," Doctor's Independent Network, accessed September 25, 2010, http://www.dinweb.org/dinweb/DINMuseum/Edward%20Jenner.asp.

91 **English doctors forced anyone awaiting variolation to endure six weeks of bleeding . . . only 850 patients had undergone the procedure:** Friedman and Friedland, *Medicine's Ten Greatest Discoveries*, 68.

91–92 **Jenner and the other children . . . even auditory hallucinations:** Ibid., 69–70.

92 **"Although changes were beginning to occur in the British medical system":** Ibid.

93 **until a cuckoo nestling was photographed in the act in 1921:** E. L. Scott, "Edward Jenner, F.R.S., and the Cuckoo," *Notes Rec. R. Soc. Lond.* 1974 28, 238 doi: 10.1098/rsnr.1974.0016.

93 **to smear Jenner long after his death:** Ibid., 235.

93 **His knife struck the clogged arteries, which he described as "bony canals":** "Jenner Centenary Number," *British Medical Journal: The Journal of the British Medical Association* 1847, May 23, 1896, 1254, accessed

September 26, 2010, http://books.google.com/books?id=hSgJAAAAIAAJ&vq=1254&pg=PA1254#v=onepage&q&f=false.

94 Jenner heard a paper presented by a Dr. Frewster: Review of *Time's Telescope for 1825*, various authors, *The Literary Gazette, and Journal of Belles Lettres, Arts, Sciences, &c.*, November 27, 1824, 757, accessed September 26, 2010, http://books.google.com/books?id=RopHAAAAYAAJ&lpg=RA1-PA757&ots=hguY5Oxhc-&dq=jenner%20mr.%20frewster&pg=RA1-PA757#v=onepage&q&f=false.

94 a Dorset gentleman farmer named Benjamin Jesty: Patrick J. Pead, "Benjamin Jesty: The First Vaccinator Revealed," *The Lancet* 368:9554, December 23, 2006, 2202, doi:10.1016/S0140-6736(06)69878-4.

94 he never developed more than a mild reaction: Friedman and Friedland, *Medicine's Ten Greatest Discoveries*, 79.

95 He chose as his test subject eight-year-old James Phipps: Peter Macinnis, interview by Robyn Williams, "Defending Edward Jenner," *Ockham's Razor*, Radio National, November 23, 1997, transcript accessed September 26, 2010, http://www.abc.net.au/rn/science/ockham/stories/s356.htm.

95 no infection developed: Riedel, "Edward Jenner and the history of smallpox and vaccination," 24.

95 "An Inquiry into the Causes and Effects of the Variolae Vaccinae": Jenner, Edward, M.D., F.R.S., *An Inquiry into the Causes and Effects of the Variolae Vaccinae, a disease discovered in some of the western counties of England, particularly Gloucestershire and Known by the Name of Cow Pox* (London: Self-published, 1800), accessed September 26, 2010, http://books.google.com/books?id=QDXShHV2z0MC&ots=L1iVu-y2GU&dq=%E2%80%9CAn%20Inquiry%20into%20the%20Causes%20and%20Effects%20of%20the%20Variolae%20Vaccinae&pg=PP11#v=onepage&q&f=false.

96 his broad enthusiasm for different branches of science: Alexandra Minna Stern and Howard Markel, "The History of Vaccines and Immunization: Familiar Patterns, New Challenges," *Health Affairs* 24:3 (2005), 611–21, doi: 10.1377/hlthaff.24.3.611.

96 has identified the "gentleman scientist": Kelty, "Meanings of Participation: Outlaw Biology?" accessed September 22, 2010, http://outlawbiology.net/about/wtf/.

97 they share a romanticism about science: Richard Holmes, *The Age of Wonder: How the Romantic Generation Discovered the Beauty and Terror of Science*, (New York: Pantheon, 2009).

98 he would enclose himself in an airtight chamber and huff quarts of newly invented laughing gas: Ibid., 235–304.

98 founder of scientific surgery . . . injected his own genitals with pus: John J. Ross, M.D., C.M., review of *The Knife Man: The Extraordinary Life and Times of John Hunter, Father of Modern Surgery*, by Wendy Moore, *New England Journal of Medicine* 353, 2005, 2412–13, accessed September 26, 2010, http://www.nejm.org/doi/full/10.1056/NEJM200512013532221.

Chapter 10: Cancer Kitchen

100 **so-called microbial fuel cells would rely on specialized bacteria:** Bruce Rittman, interviewed by Earthsky, "Bruce Rittman uses bacteria to make energy from waste," Earthsky, August 29, 2008, accessed September 26, 2010, http://earthsky.org/energy/fuel-cells-use-bacteria-to-make-energy-from-waste.

100 **the antiaging theories of de Grey:** See the SENS Foundation Web site for an extensive introduction to de Grey's theories, http://www.sens.org/sens-research.

100 **long a polarizing figure:** For a sense of the flavor of the debate over de Grey's theories, see this scathing November 29, 2005, *Technology Review* letter to the editor from University of Michigan biogerontologist Richard Miller, http://www.technologyreview.com/biomedicine/15936/page1/, and de Grey's response, http://www.technologyreview.com/biomedicine/15941/?a=f.

104 **discovered several microbes that ate their way through the cholesterol:** Jacques M. Mathieu, John Schloendorn, Bruce E. Rittmann, and Pedro J. J. Alvarez, "Medical bioremediation of age-related diseases," *Microbial Cell Factories* 8:21, 2009, doi:10.1186/1475-2859-8-21.

105 **"Biopunks deplore restrictions on independent research":** The manifesto is available in its entirety on Patterson's blog, http://maradydd.livejournal.com/496085.html.

105 **research into how the immune system might be manipulated to battle cancer:** The Web site of the International Society for Biological Therapy of Cancer is a good starting point for delving into immunotherapy approaches to battling cancer, http://www.isbtc.org/.

106 **cervical cancer cells taken from a patient named Henrietta Lacks by Johns Hopkins researchers:** Rebecca Skloot, *The Immortal Life of Henrietta Lacks* (New York: Crown, 2010).

106 **The pictures tell the story:** Livly, http://www.livly.org/Research.html.

107 **a 1957 study:** Chester M. Southam, Alice E. Moore, and Cornelius P. Rhoads, *Homotransplantation of Human Cell Lines*, Science 25, January 1957, 158–60, doi: 10.1126/science.125.3239.158.

107 **An account from *Time* written the year of the study describes what happened:** "Medicine: Cancer Volunteers," *Time*, February 25, 1957, accessed September 26, 2010, http://www.time.com/time/magazine/article/0,9171,936841,00.html.

108 **a kind of pattern recognition:** For a good online overview innate immunity, see Gene Mayer, "Innate (Non-specific) Immunity," University of South Carolina School of Medicine, accessed September 26, 2010, http://pathmicro.med.sc.edu/ghaffar/innate.htm.

109 **the U.S. Food and Drug Administration approved Provenge:** "FDA Approves a Cellular Immunotherapy for Men with Advanced Prostate Cancer," U.S. Food and Drug Administration, April 29, 2010, accessed September 26, 2010, http://www.fda.gov/NewsEvents/Newsroom/PressAnnouncements/ucm210174.htm.

109 **has made its developers millions:** John Carroll, "Dendreon CEO cashes in $28.8M of shares on Provenge OK," *FierceBiotech*, May 4, 2010, accessed September 26, 2010, http://www.fiercebiotech.com/story/dendreon-ceo-cashes-28-8m-shares-provenge-ok/2010-05-04.

109 too clever: Livly, http://www.livly.org/Research.html.

110 The Livly Web site proudly described the lab's arsenal of gear: Livly, http://www.livly.org/collaborative_Space.html.

111 Gentry began organizing meetups for Bay Area residents: BioCurious meetups, http://www.meetup.com/BioCurious/.

114 The two quickly hatched a plan to mimic Venter's self-styled epic quest to sail around the globe: See official Venter *Sorcerer II* expedition Web site, http://www.sorcerer2expedition.org/version1/HTML/main.htm.

II: Read/Write

119 biopunks would be nowhere without the Man: Kelty, "Meanings of Participation: Outlaw Biology?" accessed September 22, 2010, http://outlaw biology.net/about/wtf/

120 Cetus, one of the world's first and at the time best-financed biotech companies: Cetus's March 1981 initial public offering raised more than $100 million, at the time the largest IPO in U.S. history. Stelios Papadopoulos, "Evolving paradigms in biotech IPO valuations," *Nature Biotechnology* 19, June 2001, BE18-BE19, accessed September 26, 2010, http://www .nature.com/bioent/building/planning/012003/full/nbt0601supp_BE18 .html.

121 "Who sets the agenda? Who really innovates?": Kelty, "Meanings of Participation: Outlaw Biology?" accessed September 22, 2010, http://outlaw biology.net/about/wtf/.

Chapter 11: Reading

122 The press called him the Grim Sleeper: For the definitive account of the Grim Sleeper's crimes, see Christine Pelisek, "Grim Sleeper Returns: He's Murdering Angelenos, as Cops Hunt his DNA," *LA Weekly*, August 27, 2008, accessed September 26, 2010, http://www.laweekly.com/2008-08-28/ news/grim-sleeper/.

122 the state-run lab was comparing the DNA of a man recently convicted of a felony weapons charge: "Brown's Forensic Experts Identify Grim Sleeper Serial Killer Suspect Through Unprecedented Use of Familial DNA," California Office of the Attorney General, July 8, 2010, accessed September 26, 2010, http://ag.ca.gov/newsalerts/release.php?id=1947.

123 swiped a pizza crust: Gillian Flaccus, "Grim Sleeper DNA Trail: Pizza Crust Helped Lead Cops to Serial Killer," *The Associated Press*, July 9, 2010, accessed September 26, 2010, http://www.huffingtonpost.com/2010/07/09/ grim-sleeper-dna-trail-pi_n_641035.html.

123 dating as far back as one hundred thousand years: Ancestry.com, accessed September 26, 2010, http://dna.ancestry.com/buyKitGoals.aspx.

123 the cost of getting an entire human genome sequenced will close in on zero: Matthew Herper, "Free DNA Testing for Everyone," Forbes.com, September 15, 2010, accessed September 26, 2010, http://www.forbes .com/2010/09/15/genome-dna-mapping-2020-opinions-contributors -matthew-herper.html?boxes=Homepagechannels.

125 "Avey and Wojcicki were joined by Wendi Murdoch": Michael Schulman, "Ptooey!" *The New Yorker*, September 22, 2008, accessed September 26, 2010, http://www.newyorker.com/talk/2008/09/22/080922ta_talk_schulman.

126 For instance, in the case of alcohol flush: "Alcohol Flush Reaction," 23andme.com, accessed September 26, 2010, https://www.23andme.com/health/Alcohol-Flush-Reaction/.

127 Speculation ensued that the company's business model was out of whack: David P. Hamilton, "23andMe's Price Cut: The End of Commerical Personal Genomics?" *BNET Healthcare*, September 9, 2008, accessed September 26, 2010, http://www.bnet.com/blog/healthcare/23andmes-price-cut-the-end-of-commerical-personal-genomics/151.

127 The company endured layoffs: Jason Kincaid, "Layoffs Confirmed At 23andMe," *TechCrunch*, October 29, 2009, accessed September 26, 2010, http://techcrunch.com/2009/10/29/layoffs-confirmed-at-23andme/.

127 23andMe hosted a policy forum: "Genomics and the Consumer: The Present and Future of Personalized Medicine," 23andme.com, July 14, 2010, accessed September 26, 2010, https://www.23andme.com/policyforum/.

128 The American Civil Liberties Union objected to the bill: Turna Ray, "Privacy Groups Challenge Calif. Bill Pushing Regulatory Exemptions for 'Post-CLIA Bioinformatics Services'," *Pharmacogenomics Reporter*, June 24, 2009, accessed September 26, 2010, http://www.genomeweb.com/dxpgx/privacy-groups-challenge-calif-bill-pushing-regulatory-exemptions-post-clia-bioi?page=show.

128 announced plans to begin selling its genetic-testing kits at thousands of Walgreens' stores: Dan Vorhaus, "FDA Puts the Brakes on Pathway-Walgreens Pairing; What's Next for DTC?" *Genomics Law Report*, May 13, 2010, http://www.genomicslawreport.com/index.php/2010/05/13/fda-puts-the-brakes-on-pathway-walgreens-pairing-whats-next-for-dtc/.

128–29 Public health officials in California and New York had sent cease and desist letters: Marcus Wohlsen, "State Suspends Sales by 13 DNA Testing Start-ups," The Associated Press, June 17, 2008, accessed September 26, 2010, http://articles.sfgate.com/2008-06-17/business/17165292_1_public-health-genetic-tests.

129 California regulators did an about-face: Andrew Pollack, "California Licenses 2 Companies to Offer Gene Services," *New York Times*, August 19, 2008, accessed September 26, 2010, http://www.nytimes.com/2008/08/20/business/20gene.html.

129 the FDA sent the company an enforcement letter: James Woods, deputy director, Patient Safety and Product Quality, Office of In Vitro Diagnostic Device Evaluation and Safety, Center for Devices and Radiological Health, U.S. Food and Drug Administration, "Letter to Pathway Genomics Corporation Concerning the Pathway Genomics Genetic Health Report," U.S. Food and Drug Administration, May 10, 2010, accessed September 26, 2010, http://www.fda.gov/MedicalDevices/ResourcesforYou/Industry/ucm211866.htm.

129 In his opening statement he chided the four most prominent direct-to-consumer companies: Henry A. Waxman, "Statement of Rep. Henry A. Waxman, Chairman, Committee on Energy and Commerce, 'Direct-to-Consumer Genetic Testing and the Consequences to the Public Health' Subcommittee on Oversight and Investigations," July 22, 2010, accessed September 26, 2010, http://energycommerce.house.gov/documents/20100722/Waxman.Statement.oi.07.22.2010.pdf.

129 **Michigan Democrat Bart Stupak referred to personal genomic scans as "snake oil":** Dan Vorhaus, "'From Gulf Oil to Snake Oil': Congress Takes Aim at DTC Genetic Testing," *Genomics Law Report*, July 22, 2010, accessed on September 26, 2010, http://www.genomicslawreport.com/index.php/2010/07/22/from-gulf-oil-to-snake-oil-congress-takes-aim-at-dtc-genetic-testing/.

129 **the GAO revealed that investigators had conducted a yearlong investigation:** Gregory Kutz, managing director, Forensic Audits and Special Investigations, U.S. Government Accountability Office, "Direct-to-Consumer Genetic Tests: Misleading Test Results Are Further Complicated by Deceptive Marketing and Other Questionable Practices," hearing on "Direct-to-Consumer Genetic Testing and the Consequences to the Public Health" before the House Committee on Energy and Commerce, Subcommittee on Oversight and Investigations, July 22, 2010, accessed September 26, 2010, http://energycommerce.house.gov/documents/20100722/Kutz.Testimony.07.22.2010.pdf.

130 **Gregory Kutz played a YouTube video of recorded phone calls:** Hearing on "Direct-to-Consumer Genetic Testing and the Consequences to the Public Health," accessed September 26, 2010, http://www.youtube.com/watch?v=ngdRUoPAQM0&feature=player_embedded.

130 **"the most accurate way for these companies to predict disease risks":** Gregory Kutz, "Direct-to-Consumer Genetic Tests: Misleading Test Results Are Further Complicated by Deceptive Marketing and Other Questionable Practices," 9.

130 **Supporters of direct-to-consumer genetic testing services saw the hearings as a dark day:** Daniel MacArthur, "A sad day for personal genomics," Genomes Unzipped, July 22, 2010, accessed September 26, 2010, http://www.genomesunzipped.org/2010/07/a-sad-day-for-personal-genomics.php.

131 **"we are troubled and find the report is deeply flawed":** Entry in The Spittoon blog, "GAO Studies Science Non-Scientifically," July 23, 2010, accessed November 1, 2010, http://spittoon.23andme.com/2010/07/23/gao-studies-science-non-scientifically/.

131 **"We just don't know how people will use the information":** Marcus Wohlsen, "Bipolar Disorder At-Home Test Causes Stir," The Associated Press, March 22, 2008, accessed September 26, 2010, http://www.huffingtonpost.com/2008/03/24/bipolar-disorder-athome-t_n_93087.html.

132 **Oz said that the results meant he would not have to subject himself to the unpleasant regular prostate screenings:** Kevin Davies, *The $1,000 Genome: The Revolution in DNA Sequencing and the New Era of Personalized Medicine* (New York: Free Press, 2010), 164, accessed September 26, 2010, http://books.google.com/books?id=dY7zDVhNYQkC&lpg=PA164&ots=7KL5t7okQu&dq=mehmet%20oz%2023andme%20prostate&pg=PA164#v=onepage&q=mehmet%20oz%2023andme%20prostate&f=false.

132 **"I have little patience for the argument that we need doctors as gatekeepers":** Thomas Goetz, "Why the Debate Over Personal Genomics Is a False One," *The Decision Tree*, May 21, 2010, accessed on September 26, 2010, http://thedecisiontree.com/blog/2010/05/why-the-debate-over-personal-genomics-is-a-false-one/.

135 millions of animal species: Paul D. N. Hebert, Sujeevan Ratnasingham, and Jeremy R. de Waard, "Barcoding Animal Life: Cytochrome C Oxidase Subunit 1 Divergences Among Closely Related Species," *Proceedings of the Royal Society of London, B,* 2003 270, S96-S99 doi: 10.1098/rsbl.2003.0025.

135 The Cancer Genome Atlas: http://cancergenome.nih.gov/.

138 mutations in a gene known as the MTHFR gene: "MTHFR," *Genetics Home Reference: Your Guide to Understanding Genetic Conditions,* U.S. National Library of Medicine, National Institutes of Health, last updated September 19, 2010, accessed September 26, 2010, http://ghr.nlm.nih.gov/gene/MTHFR.

140 crossover study: Bonnie Sibbald and Chris Roberts, "Understanding controlled trials: Crossover trials," *British Medical Journal* 316:7146, June 6, 1998, 1719, accessed September 26, 2010, http://www.bmj.com/content/316/7146/1719.full.

141 a Web site called SNPedia: http://www.snpedia.com/index.php/SNPedia.

141 A free program called Promethease: http://www.snpedia.com/index.php/Promethease.

142 a DIY genomics smartphone app: http://www.diygenomics.org/index.php.

Chapter 12: Writing

143 "unbelievable low prices": Mr. Gene GmbH, accessed September 26, 2010, http://mrgene.com/desktopdefault.aspx/tabid-2/.

143 thirty-nine cents a letter: "DNA best pricing," Mr. Gene GmbH, accessed September 26, 2010, http://mrgene.com/desktopdefault.aspx/tabid-77/.

144 the idea of the "long tail": Chris Anderson, "Long Tail 101," *The Long Tail,* September 8, 2005, accessed September 26, 2010, http://www.longtail.com/the_long_tail/faq/.

144 Anderson's lab at the University of California, Berkeley, occupies a corner of the campus's grand new life-sciences building: For more on Anderson and his work, see his lab's Web site, http://andersonlab.qb3.berkeley.edu/.

146 Understanding protein folding is one of the most challenging problems in modern molecular biology: For a really cool introduction to protein folding and a crowdsourced effort to understand its intricacies, see Stanford's "Folding@home" project, http://folding.stanford.edu/.

148 his synthetic biology software, called Clotho: Download Clotho at http://clothocad.org/.

150 the device, called a CoolBot: Store It Cold LLC, http://www.storeitcold.com/.

151 In 2008, a team from Rice University developed bacteria to brew beer that contained resveratrol: Rice University Biobeer Project, http://2008.igem.org/Team:Rice_University.

151 students from Slovenia took first prize for engineering a vaccine for the microbe that causes stomach ulcers: Alla Katsnelson, "Synthetic vaccine nabs iGEM prize," *The Scientist,* November 10, 2008, accessed September 26, 2010, http://www.the-scientist.com/blog/display/55178/.

152 Berkeley's 2010 iGEM entry: For more details on Berkeley 2010 iGEM entry, see IGEM:Berkeley/2010, http://openwetware.org/wiki/IGEM:Berkeley/2010.

153 that all contain the same complementary sequences of letters

at each end: For a brief technical explanation of how BioBrick parts fit together, see the description of the BioBrick Assembly Kit from Gingko Bioworks, http://ginkgobioworks.com/biobrickassemblykit.html.

155 **"the rottenest city":** "Emeryville Is Born–1890s to 1930s," City of Emeryville Web site, accessed September 26, 2010, http://www.emeryville.org/index.aspx?NID=660.

155 **The BIOFAB was founded in late 2009 as the world's first open-source machine shop for synthetic biology:** "About the BIOFAB," BIOFAB Web site, http://www.biofab.org/about.

156 **the bacteria's genome ran about 1.1 million letters long:** "First Self-Replicating, Synthetic Bacterial Cell Constructed by J. Craig Venter Institute Researchers," J. Craig Venter Institute, May 20, 2010, accessed September 26, 2010, http://www.jcvi.org/cms/press/press-releases/full-text/article/first-self-replicating-synthetic-bacterial-cell-constructed-by-j-craig-venter-institute-researcher/.

158 **480 inches:** "Cherrapunji, India: Climate, Global Warming, and Daylight Charts and Data," Climate-Charts.com, accessed September 26, 2010, http://www.climate-charts.com/Locations/i/IN42515.php.

158 **In Cherrapunji, the bridges are alive:** For a series of fantastic photos and a detailed account of the bridges of Cherrapunji, see *Living Root Bridges*, http://rootbridges.blogspot.com/.

159 **figured a way to use that the same genetic machinery researchers are working with to create fuels to make artemisinin:** Michael Specter, "A Life of Its Own: Where Will Synthetic Biology Lead Us?" *The New Yorker*, September 28, 2009, accessed September 26, 2010, http://www.michaelspecter.com/2009/09/new-yorker-article-number-two/.

159 **some of the world's poorest people also depend for a living on making and selling artemisinin:** Jim Thomas, program manager, ETC Group, "Benefits and Risks of Synthetic Biology," testimony before the Presidential Commission for the Study of Bioethical Issues, July 8, 2010, transcript accessed September 26, 2010, http://www.bioethics.gov/transcripts/synthetic-biology/070810/benefits-and-risks-of-synthetic-biology.html.

III: Safety/Risk

167 **Shortly before the exhibit was to open, Hope Kurtz collapsed from heart failure:** This account of Kurtz's case is drawn from multiple sources, including: "Frequently Asked Questions," Critical Art Ensemble Defense Fund, http://www.caedefensefund.org/faq.html; "Charge Dropped Against Artist in Terror Case," The Associated Press, April 22, 2008, accessed September 26, 2010, http://www.nytimes.com/2008/04/22/nyregion/22bioart.html?_r=1; and Lynne Duke, "The FBI's Art Attack: Offbeat Materials at Professor's Home Set Off Bioterror Alarm," *Washington Post*, June 2, 2004, accessed September 26, 2010, http://www.washingtonpost.com/wp-dyn/articles/A8278-2004Jun1.html.

Chapter 13: Threat

171 **Aafia Siddiqui was born into an upper-middle-class Karachi family in 1972:** Deborah Scroggins, "The Most Wanted Woman in the World," *Vogue*, March 2005.

171 **While in school, she allegedly raised money for charities with ties to Islamist extremists:** Ibid., 266.

171 **now claims U.S. authorities held her for years in a secret prison:** "Press Release: Aafia Siddiqui Claims She Was Held by the U.S. in Bagram For Years," freedetainees.org, August 7, 2008, accessed September 26, 2010, http://freedetainees.org/1672.

171 **a federal grand jury indictment says that when she was picked up she was carrying handwritten notes:** *United States of America v. Aafia Siddiqui* (S.D. New York 2008), accessed September 26, 2010, http://www.nefafoundation.org/miscellaneous/FeaturedDocs/US_v_Siddiqui_ind.pdf.

172 **a U.S. District Court jury in Manhattan convicted Siddiqui:** C. J. Hughes, "Pakistani Scientist Found Guilty of Shootings," *New York Times*, February 3, 2010, accessed September 26, 2010, http://www.nytimes.com/2010/02/04/nyregion/04siddiqui.html.

172 **Siddiqui took the stand:** C. J. Hughes, "Neuroscientist Denies Trying to Kill Americans," *New York Times*, January 28, 2010, accessed September 26, 2010, http://www.nytimes.com/2010/01/29/nyregion/29siddiqui.html.

172 **During the sentencing hearing, Siddiqui disputed she was ill:** Benjamin Weiser, "Pakistani Sentenced to 86 Years for Attack," *New York Times*, September 23, 2010, accessed September 26, 2010, http://www.nytimes.com/2010/09/24/nyregion/24siddiqui.html.

172 **pointed to Siddiqui as exactly the kind:** Bob Graham and Jim Talent, et al., *World at Risk: The Report of the Commission on the Prevention of WMD Proliferation and Terrorism*, December 2008, accessed September 22, 2010, http://www.preventwmd.gov/world_at_risk_biological_and_nuclear_risks/.

172 **"biotechnology has spread globally":** Ibid., http://www.preventwmd.gov/world_at_risk_preface.

173 **"The nuclear age began with a mushroom cloud":** Ibid.

173 **"What's available to idealistic students":** Michael Specter, "A Life of Its Own: Where Will Synthetic Biology Lead Us?" *The New Yorker*, accessed September 26, 2010, http://www.michaelspecter.com/2009/09/new-yorker-article-number-two/

173 **"The ability to create nasty pathogens":** Anthony L. Kimery, "Zombies, Rabies and Synthetic Genomics," *HSToday*, June 2, 2009, accessed September 27, 2010, http://www.hstoday.us/content/view/8747/150/.

175 **Walt pulls a small plastic baggie from his pocket:** Bryan Cranston, director, "Seven-thirty-seven," *Breaking Bad*, Season 2, Episode 1, AMC network.

176 **the FBI said the perpetrator:** Eric Lichtblau and Megan Garvey, "Loner Likely Sent Anthrax, FBI Says," *Los Angeles Times*, November 10, 2001, accessed September 27, 2010, http://articles.latimes.com/2001/nov/10/news/mn-2459.

176 **He died four days after the attack:** Jonathan Brown, "Poison umbrella murder case is reopened," *The Independent*, June 20, 2008, accessed September 27, 2010, http://www.independent.co.uk/news/uk/crime/poison-umbrella-murder-case-is-reopened-851022.html.

176 **Bergendorff pleaded guilty to possessing a biological toxin:** Steve Friess, "In Accord, Ricin Owner Enters Plea of Guilty," *New York Times*,

August 5, 2008, accessed on September 27, 2010, http://www.nytimes.com/2008/08/05/us/05ricin.html?_r=2&ref=ricin_poison.

176 with the section on ricin production highlighted: Abigail Goldman, "Meet the Mysterious Roger Von Bergendorff," *Las Vegas Sun*, March 5, 2008, accessed September 27, 2010, http://www.lasvegassun.com/news/2008/mar/05/meet-mysterious-roger-von-bergendorff/.

177 $190,000 in debts reported in a bankruptcy filing: Ibid.

177 the smallest free-living organisms are bacteria that contain about six hundred thousand letters: "From the Genome to the Proteome: Basic Science," *The Science Behind the Human Genome Project*, U.S. Department of Energy Office of Science, accessed September 27, 2010, http://www.ornl.gov/sci/techresources/Human_Genome/project/info.shtml.

177 In testimony before a House Committee on Energy and Commerce hearing on synthetic biology in May 2010: Drew Endy, "Testimony to the House Committee on Energy and Commerce on Advances in Synthetic Biology and Their Potential Impact," May 27, 2010, accessed September 27, 2010, http://energycommerce.house.gov/documents/20100527/Endy.Testimony.05.27.2010.pdf.

179 An innocuous post to the DIYbio mailing list seeking suggestions for presentations quickly turned into an electronic slugfest: See the entire exchange on the DIYbio Google group, starting with the Chris Kelty's original post, "Outlaw Biology Symposium and Faire, UCLA, Jan 29–30, Call for Ideas," October 21, 2009, http://groups.google.com/group/diybio/browse_thread/thread/f753ea0e4685dc9d/e19bedc9b5bdd82e.

180 "You guys have known me long enough to know that I am a contrary bastard": Patterson's post is part of a follow-up exchange, "This is getting silly," started October 26, 2009, http://groups.google.com/group/diybio/browse_thread/thread/5fad85c21d0a6af0/72b1ed547a8ab14b?lnk=gst&q=contrary+bastard#72b1ed547a8ab14b.

180 "It was outreach, not oversight": Edward H. You, supervisory special agent, Federal Bureau of Investigation, "Federal Oversight of Synthetic Biology," testimony before the Presidential Commission for the Study of Bioethical Issues, July 9, 2010, transcript accessed September 27, 2010, http://www.bioethics.gov/transcripts/synthetic-biology/070910/federal-oversight-of-synthetic-biology.html.

Chapter 14: Outbreak

184 In *Engines of Creation*, MIT engineering grad Eric Drexler described the coming nanotechnology revolution: K. Eric Drexler, *Engines of Creation: The Coming Era of Nanotechnology* (New York: Anchor Books, 1986), free HTML edition, http://e-drexler.com/d/06/00/EOC/EOC_Table_of_Contents.html.

185 "You have an actual living, reproducing machine; it's nanotechnology that works": Drew Endy, "Engineering Biology: A Talk with Drew Endy," *Edge: The Third Culture*, accessed September 27, 2010, http://www.edge.org/3rd_culture/endy08/endy08_index.html.

185 Drexler himself later disavowed: Eric Drexler interviewed by nanotechweb.org, "Drexler Dubs 'Gray Goo' Fears Obsolete," June 9, 2004, accessed September 27, 2010, http://nanotechweb.org/cws/article/indepth/19648.

185 **the ETC Group called for a global moratorium on synthetic biology. . . . "We know . . . that their use threatens existing natural biodiversity":** "Synthia is Alive . . . and Breeding: Panacea or Pandora's Box?" ETC Group press release, May 20, 2010, accessed September 27, 2010, http://www.etcgroup.org/en/node/5142.

187 **Against that backdrop, Thomas's warning was stark:** Jim Thomas, programme manager, ETC Group, "Benefits and Risks of Synthetic Biology," testimony before the Presidential Commission for the Study of Bioethical Issues, July 8, 2010, http://www.bioethics.gov/transcripts/synthetic-biology/070810/benefits-and-risks-of-synthetic-biology.html.

188 **"Powerful technology in an unjust world is likely to exacerbate the injustice":** Jim Thomas, "Synthetic Biology Debate," debate with Drew Endy, The Long Now Foundation, San Francisco, November 17, 2008, http://www.longnow.org/seminars/02008/nov/17/synthetic-biology-debate/.

188 **what he calls synthetic chemistry:** Thomas, "Benefits and Risks of Synthetic Biology," http://www.bioethics.gov/transcripts/synthetic-biology/070810/benefits-and-risks-of-synthetic-biology.html.

188 **Frederick Wöhler combined two inorganic chemicals to create urea:** E. Kinne-Saffran and R. K. H. Kinne, "Vitalism and Synthesis of Urea from Friedrich Wöhler to Hans A. Krebs," *American Journal of Nephrology* 19, 1999, 290–94, doi: 10.1159/000013463.

188–89 **"Are synthetic chemists playing God? . . . just as cautionary voices on biotech are attacked today":** Thomas, "Benefits and Risks of Synthetic Biology," http://www.bioethics.gov/transcripts/synthetic-biology/070810/benefits-and-risks-of-synthetic-biology.html.

189 **Exxon's $600 million deal with Synthetic Genomics:** "ExxonMobil to Launch Biofuels Program," ExxonMobil press release, July 14, 2009, accessed September 27, 2010, http://www.businesswire.com/portal/site/exxonmobil/index.jsp?ndmViewId=news_view&ndmConfigId=1001106&newsId=20090714005554&newsLang=en.

189 **Venter's company is working to engineer the algae to constantly secrete the oil straight out of its cell walls:** "Next Generation Fuels & Chemicals," Synthetic Genomics, http://www.syntheticgenomics.com/what/renewablefuels.html.

191 **"Worker safety cannot be sacrificed on the altar of innovation":** Andrew Pollack and Duff Wilson, "Safety Rules Can't Keep Up with Biotech Industry," *New York Times*, May 27, 2010, accessed September 27, 2010, http://www.nytimes.com/2010/05/28/business/28hazard.html.

191 **thirty-one-year-old British scientist Jeannette Adu-Bobie arrived in the country in 2005 to study meningococcal bacteria:** "Amputee Scientist Infected at Lab: I Can Move On Now," *Dominon Post*, March 8, 2008, accessed September 27, 2010, http://www.stuff.co.nz/national/562741.

191 **must have coincidentally contracted the rare disease somewhere else:** "ESR 'Stuck to Guns' Over Lab Bug Denial," *Dominion Post*, April 8, 2008, accessed September 27, 2010, http://www.stuff.co.nz/dominion-post/national/564193.

191 **the New Zealand government's labor department reversed its original conclusion. . . . "There is no compelling evidence that this infection was contracted anywhere else":** Dr. Geraint Emrys, "Report

into the Investigation of ESR Meningitis Infection Case of Dr. Jeannette Adu-Bobie," New Zealand Department of Labor, July 30, 2008, accessed September 27, 2010, http://www.dol.govt.nz/News/Media/2008/adu-bobie-report.asp.

IV: Life/Science

195 **"looking for problems where elegant mathematics could be usefully applied":** "About Freeman Dyson," Freeman J. Dyson's homepage, http://www.sns.ias.edu/~dyson/about.html.

195 **the world's most impeccably credentialed global warming skeptic:** Nicholas Dawidoff, "The Civil Heretic," *New York Times Magazine*, March 25, 2009, accessed September 27, 2010, http://www.nytimes.com/2009/03/29/magazine/29Dyson-t.html.

195 **"Our Biotech Future," an article Dyson wrote for *The New York Review of Books* in 2007:** Dyson, "Our Biotech Future," *The New York Review of Books*, July 19, 2007, http://www.nybooks.com/articles/archives/2007/jul/19/our-biotech-future.

196 **Wendell Berry called Dyson the latest in a line of soothsayers:** Wendell Berry, James P. Herman, and Christopher B. Michael, reply by Freeman Dyson, " 'Our Biotech Future': An Exchange," *The New York Review of Books*, September 27, 2007, accessed September 27, 2010, http://www.nybooks.com/articles/archives/2007/sep/27/our-biotech-future-an-exchange/.

198 **Myers wrote in a 2010 riposte:** P. Z. Myers, entry posted to Pharyngula blog, "Ray Kurzweil does not understand the brain," August 17, 2010, accessed November 1, 2010, http://scienceblogs.com/pharyngula/2010/08/ray_kurzweil_does_not_understa.php.

201 ***GFP Bunny* by globe-trotting bioartist Eduardo Kac:** Eduardo Kac, "GFP Bunny," artist's Web site, http://www.ekac.org/gfpbunny.html.

201 **Imagine a project in which living cells are draped on a frame and kept alive using fetal calf serum to grow into a jacket:** Frances Stracey, "Bio-art: The Ethics Behind the Aesthetics," *Nature Reviews Molecular Cell Biology* 10, 496–500, July 2009, doi:10.1038/nrm2699.

202 **San Franicsco Bay Area bioartist Philip Ross:** See Ross's personal Web site for images and explanations of his work, http://www.philross.org/.

207 **Backyard Brains:** http://www.backyardbrains.com/.

209 **Watt once described punk this way:** Karen Schoemer, "Mike Watt Bio," Mike Watt's Hoot Page, October 2005, accessed September 27, 2010, http://www.hootpage.com/hoot_wattbio.html#wattbiokschoemer2005.

Index